ASPECTS OF A STRATIGRAPHIC SYSTEM THE DEVONIAN

Aspects of a Stratigraphic System: the Devonian

D. L. DINELEY

Department of Geology
University of Bristol
England

A HALSTED PRESS BOOK

JOHN WILEY & SONS
New York

First published 1984 by
Higher and Further Education Division
MACMILLAN PUBLISHERS LTD
London and Basingstoke

Published in the U.S.A. by
Halsted Press, a division of
John Wiley & Sons, Inc., New York

Printed in Hong Kong

Library of Congress Cataloging in Publication Data

Dineley, David L.
   Aspects of a stratigraphic system — The Devonian.

   "A Halsted Press book."
   Includes bibliographical references and index.
   1. Geology, Stratigraphical — Devonian.   I. Title.
QE665.D56   1984   551.7′4   84–5266
ISBN 0–470–20086–3

Library of Congress Catalog Card No.: 84–5266

Contents

v

Preface

The beautiful county of Devon has lent its name to an important and widespread stratigraphic system. The title was proposed by two of the most active and widely experienced geologists in Victorian Britain, men who also influenced numbers of like-minded scientists in Europe, North America and elsewhere. Sedgwick and Murchison would, no doubt, be pleased to see how the sciences of palaeontology and stratigraphy have prospered since 1839 when the term Devonian was given a stratigraphic meaning. They would also perhaps, with their truly Victorian firmly held views and sense of purpose, be eager to debate the relevance of these advances to the newer fields of sedimentological, palaeomagnetic and tectonic studies (and vice versa). They would appreciate the many aspects that a stratigraphic system has and how these change while science advances.

The geological sciences indeed advance rapidly on all fronts and our knowledge of the world as it was in the distant past increases at an ever quickening pace. Much of the new information soon finds its way into lecture courses and textbooks and these tend to follow a pattern that is recognisably conservative and rather predictable. The present volume was prompted not by the wish (or need) to produce an exhaustive or reference treatise on the Devonian system for the scholar or researcher, nor by the desire to produce yet another textbook. It seeks to illustrate the ways in which the different disciplines or interests within geology assist in a general understanding of how things may have been 350 million years ago or thereabouts. It is intended as a book for the interested but not very advanced student of historical geology, and, unintentionally, it no doubt illustrates the difficulties that the writer familiar with one aspect of geology has when he has to try to understand the importance another aspect may have for a theme as big as this one.

A practice among scientists is to prefer the simplest explanation for a collection of facts or events—usually this turns out to be inadequate and more complicated explanations are called for. Thus almost certainly the models of the Devonian world and its inhabitants are naive and even inappropriate. Each of the general subject areas covered in the following chapters makes a contribution to the description and assessment of what took place far back in a specific period of geological time. Each is of importance to one or more of the others and each can be dealt with only briefly here. Some fields of study have been completely neglected in what follows: Devonian volcanism is an example. Others will become apparent.

Comparisons with the state of knowledge and interpretations of other geological systems may be obvious, but the Devonian is nevertheless important in several unique aspects. It was a period of profound, even violent, geological and geographical events on the surface of the earth. Plant and animal life responded vigorously. Terrestrial vegetation evolved rapidly to cover more of the land surface than ever before and the vertebrates followed a conquest of both salt and sweet waters by crawling off into the undergrowth. There were also great extinctions and changes in the marine invertebrate world and we are still far from explaining all of these events.

The need for a proper understanding of the way the earth works, and worked then, is paramount and, for the scientifically curious, the 50 million years of the Devonian period are as good as any against which to test new ideas. This account must be brief; it is of necessity incomplete and tentative.

Several points might be nevertheless emphasised here. Palaeomagnetic studies seem to abound in inconsistencies. Those of Devonian rocks are no exception and the accuracies of palaeogeographic reconstructions probably suffer because of possible palaeomagnetic overprinting on the widely distributed Devonian outcrops throughout subsequent geological ages. Palaeomagnetic reversals, similarly, do not seem as yet to offer much help in Devonian correlation. Debate about the extent to which plate-tectonic theory satisfactorily explains all Palaeozoic palaeogeography still goes on. The pangeic view of the Palaeozoic (as distinct from the 'wandering continent' or kaleidoscopic view) has its ardent adherents and the 'expanding Earthers' are not silenced. In the ensuing pages an essentially orthodox 'steady-state' plate-tectonic view of the earth during Phanerozoic time is maintained. Problems about Devonian faunal and floral distributions remain despite the means of dispersal and isolation being more convincingly depicted in recent publications. The remarkable rise and spread of the vertebrates that are normally regarded as typical of non-marine Old Red Sandstone areas calls for further study. And withal it must be remembered that the Devonian period of some 50 million years lasted almost as long as has the Cainozoic era. The extent of the movements of continents and the extent of changes in the living world in the Devonian were probably not very different from those of the Cainozoic. And both may be called an 'Age of Fishes' though the differences are fairly obvious!

My early contacts with Devonian studies were much encouraged by Dr E. I. White and by the late Professor L. J. Wills. I have since been fortunate in many friends who have helped, by discussion and encouragement, in the writing of this book and I would like to thank them all. In particular my gratitude goes to Professor T. S. Westoll, P. F. Friend, H. G. Reading, E. J. Loeffler, K. C. Allen, B. P. J. Williams, H. H. Tsien, Liu Yuhai, Hou Hong-fei and J. W. Cowie. I would also like to record thanks to members of the I.C.S. Subcommission on the Devonian System who have been so helpful on numerous occasions. Mrs J. D. Rowland turned my manuscript into typescript and Mrs A. Gregory provided the draughting from my sketches—all to my great satisfaction. The blemishes that remain are mine alone.

D.L.D.
Bristol, 1983

1

Introduction

Each period of geological time has had its special attributes and events; each has left a distinct and varied record that tells us something of the evolution of our planet. Deciphering the record is one of the main aims of geology, an aim that is more difficult to achieve the further back in geological time we search. Subsequent events mask the nature of the original record but the stratigraphical use of fossils, palaeomagnetism and the application of radiometric dating are increasingly useful in enabling us to ascribe relative and numerical dates to geological phenomena. The record is overall imperfect but continuous, since constructive and destructive processes are operating ceaselessly upon the earth's crust. This represents a hugely dynamic state of affairs and the continuous expenditure of energy. Such is the nature of the crust that it is only in the continental parts that events more than 200 m.y.b.p. or so are recorded. The oceanic realm of earlier days remains more conjectural than does the continental.

Geology today stresses that the size, shape and disposition of the continents are largely due to movements within the mantle and the slow but inexorable shifting of the crustal plates. Knowledge of how long the crustal plate-shifting processes have been operating is decidedly limited, and the processes must themselves be an expression of the evolution of the planet throughout time. The internal reorganisation or development of earth is still proceeding—its internal heat is far from exhausted. At times during the evolution of the interior no doubt strong repercussions of internal changes were manifested at the surface—in the oceans and atmosphere as well as in the solid ground.

Even within the Phanerozoic eon we have cause to wonder how far we can safely apply the concept of uniformitarianism. The explanations of continental changes, of the rise and fall of sea level and other phenomena, excite the imagination and the models now offered are more convincing than was previously the case. Proof of their absolute validity eludes us so far. Changes in the organic world and the causes of the great extinctions are equally puzzling and new concepts about them are being eagerly debated. The new evidence of anoxic periods in the oceans prompts us to reconsider the models that we have for the evolution and interactions of the hydrosphere, atmosphere and biosphere.

Prior to a consideration of the big problems we should no doubt attend to some of the smaller ones. Without understanding the Recent it is very difficult to propose convincing explanations of the Precambrian. The Palaeozoic world is, so to speak, suspended between these two extremes. It was during that time that the earth's surface would have begun to look much as we know it today but with some important and striking differences (see figure 1.1). The great geological processes, plate behaviour etc. probably functioned much as they do today. Perhaps we can apply a degree of uniformitarianism with some confidence; perhaps we can also call upon catastrophes, sudden and 'traumatic' biological changes and other rare events to explain what the record seems to hold.

Stratigraphy remains the basic source of information on which to base models of ancient environments, palaeogeographies and, hence, in large measure the history of the surface of the earth. Tectonics, palaeontology and sedimentology, geophysics and geochemistry all provide means of improving the quality of our palaeogeographical reconstructions and models of past environments.

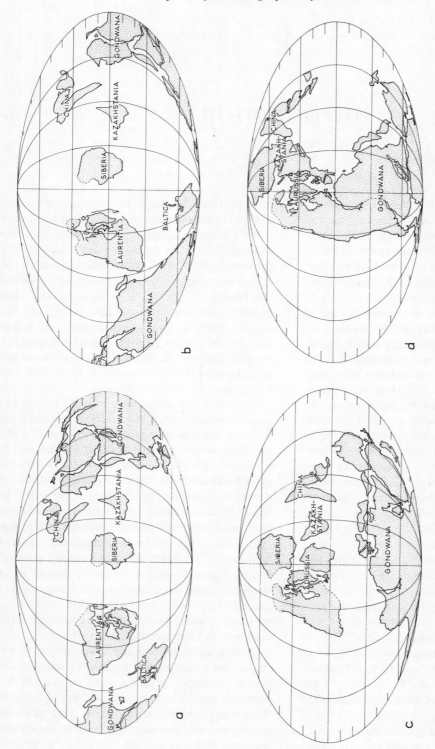

Figure 1.1. World palaeogeographic maps (after Ziegler, A. M. *et al.*, 1979). (a) Late Cambrian (Franconian); (b) Middle Ordovician (Llandilian–earliest Caradocian); (c) Early Devonian (Emsian); (d) Late Permian (Kazanian). Drawn on the Mollweide projection, these maps illustrate the major 'continental' or cratonic elements of the globe throughout Palaeozoic times. They also show the changing distribution of the 'continents' from broadly scattered through the latitudinal state in Devonian time to the more longitudinally extended 'Pangaea' at the end of the Palaeozoic era. The stratigraphy of each successive period has been in part controlled by the distribution and relief of the land masses and by their attendant climates. The question of locating the margins of ancient crustal plates remains dependent upon stratigraphy, tectonics and allied studies. Parts of some subducting margins may be located relatively easily, but the sites of (growing) oceanic ridge margins can only be guessed.

Nevertheless, geological history remains a tantalising affair in which we have a glimpse of the advertising 'stills' rather than the full and action-packed 'movie' itself.

Earlier ideas about the changing shapes and positions of the continents were discarded by many geologists when Wegener's short but epoch-making *Origins of Continents and Oceans* was published. Geophysicists remained hostile to the concept of continental drift until, ironically, studies of palaeomagnetism and the geophysical characteristics of ocean basins led to hypotheses of ocean-floor spreading and plate tectonics. Since the first plate-tectonic models of the 1960s an unending flow of papers has attempted to assess stratigraphic successions, and the evolution of the environments that they represent, in this new light. By no means all are successful and it is recognised that the model to be adopted represents a phenomenon that has itself evolved with geological time. It is now generally assumed that the *basic* plate-tectonic process and its global significance seem not to have changed greatly during Phanerozoic time. Even granting this, it is difficult to explain many ancient tectonic events and palaeogeographical changes. The Hercynian orogeny in Europe remains in want of a fully acceptable plate-tectonic explanation despite the great wealth of knowledge that we possess about the rocks involved. A convincing model will have to show the means by which the complex post-Caledonian stratigraphy arose and how its rocks were then deformed. To put it another way, the model will have to stand upon a firm foundation of *stratigraphical* as well as structural data.

Hallam (1978) has listed ten topics of major scientific importance for which a knowledge of stratigraphy is essential:

1. The nature of formerly widespread epicontinental seas.
2. Eustatic changes of sea level.
3. Ancient ocean current systems.
4. Environmental controls on organic distributions in the past.
5. Organic extinction and evolution in relation to changing environments.
6. Changing patterns of sedimentation through time.
7. Changing palaeoclimates.
8. The character of orogeny in space and time.
9. Uplift and subsidence of continental margins and interiors.
10. Relationship between orogeny and epeirogeny.

These topics, as they relate to a particular period of geological time, are in part the subject of this book. That we have still far to go in recognising and organising the data upon which a proper understanding should be based will be obvious. However, it is the journey towards that goal that continues to be so attractive to the stratigrapher. House (1975a, 1975b), when considering facies and correlation problems in Devonian strata, has posed the three prime questions for the European stratigrapher. Is there any pattern in the timing of major periods of transgression and regression around the Baltic–Laurentian shield area which in Devonian times marked the Old Red Sandstone Continent? Can these facies movements be related to tectonic events expressed by either igneous or orogenic events? Can the pattern that results be interpreted in a model that is consistent with the present-day plate-tectonic model?

The study of the geological past, then, requires more than simple stratigraphy. If we wish to examine in detail one particular geological period we need to know something of what went before and what came after it and to understand the processes that brought about the changes. The overall picture to be gained might tell us not only more about the particular span of time that first attracted our attention, but also more about earth's anatomy and evolution generally and the condition of the living world before, during and after it. As a vehicle by which to demonstrate general principles of geology, the study of a particular geological system of rocks and the phase of earth history it represents seems to be a good one. Can we see relevance for this study in the new ideas about the solar system, the possible changing rate of seafloor spreading, geological cyclicity, provinciality among the marine benthos, and so on? The so-called first law of ecology, 'that everything is linked to everything else', holds good for geology too; we must always probe relationships. This book only airs a few of the more obvious and attractive problems that arise in a study of a particular period of not-too-far distant earth history. A duration of

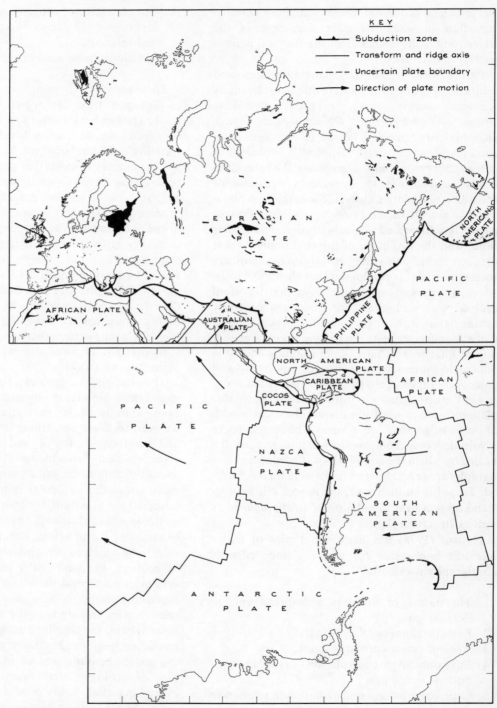

Figure 1.2. Outcrops of Devonian rocks are shown on a world map on which present-day underlying cause of not only continental drift but also much volcanism, sea-level changes and different global pattern of plates, but the exact configuration of the plates is largely obscure largely of epicontinental sediments that were deposited in basins on rigid lithosphere. Many with mobile belts, megasutures or megashear systems (see Bally and Snelsdon, 1980).

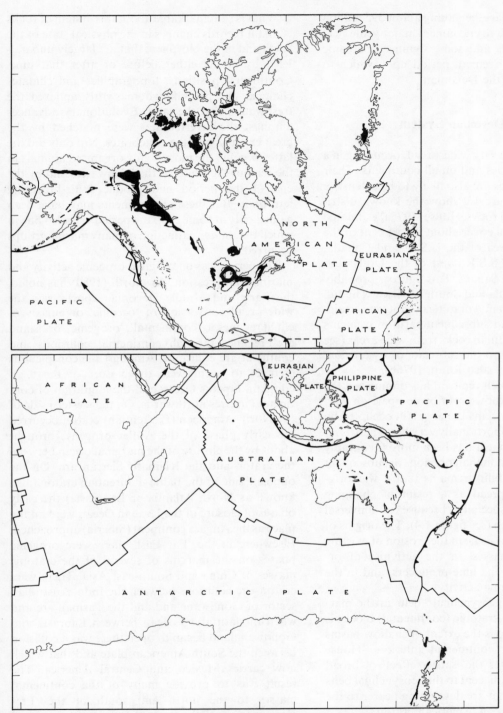

plate boundaries are suggested. The evolution of the plates themselves is now seen as the
other influences upon the stratigraphic record. Devonian rocks originated upon a very
despite the reconstruction that is possible, as shown in figure 1.1. The broader outcrops are
of the smaller linear outcrops represent formations that were deposited in basins associated

about 50 million years, beginning about 395 million years ago, is enough to encompass major geological events and changes and some significant organic changes too. A convenient period upon which to base our debate is the Devonian.

The Devonian System

Devonian rocks are very widespread, occurring in a great variety of facies and on all continents. Their offshore occurrences are also likely to be extensive.

The maps in figure 1.2 show the known distribution of Devonian rocks. House (1975a) estimated that these rocks still cover about 29 per cent of the continental land area of the U.S.A. and Canada, Europe and the U.S.S.R. west of the Urals, and Africa north of the Sahara. This figure is probably exceeded in Australia and South America. The full extent of the system in Antarctica is not yet certain. Late Palaeozoic and subsequent earth movements have removed Devonian rocks from wide areas (see figure 1.3), either into the substrate or by erosion, from the record (see also Raup, 1976b).

The extent to which removal has diminished the record remains unknown, but by comparison with other Palaeozoic systems the area still occupied by the Devonian is proportionally very big. There is no reason to believe that it has been more favourably protected from erosion than previous or subsequent systems: the probability remains that it was originally more widespread as a result of changing geography and, especially, because of universal marine transgression (see figure 1.4). The origins of the deposited materials lie in the erosion of existing and rising land masses, in the activities of organisms, especially the lime-producers, and in the magnitude of volcanic activity.

The presence of geosynclinal* belts at the margins of many of the cratonic (continental) blocks is conspicuous, as also is the extent of shallow basins that cover several continental interiors. House (1975a, p. 238) notes the 'saucer effect' of broad continental uplifts adjacent to the geosynclinal belts and broad basins that are developing nearer to the centre of the European–American craton.

---

* Although the term 'geosyncline' has fallen from favour, it may still be used in the broad sense of a long-lived tectonic or mobile belt wherein thick sediments accumulate.

What is so remarkable about this system of rocks is that it records changes in the physical state of the earth and in the biosphere that are largely unparalleled by events either before or after that time. Continental positions, topographies and climates changed; the plant kingdom swiftly annexed the freshwaters and the land. Evolutionary advances amongst the invertebrates were matched by the great expansion of the vertebrates. Not only did the latter become the most highly organised animals in the sea but they entered the freshwaters and by the end of the period had produced air-breathing tetrapods. But these achievements may have been gained only at some cost: the period was also one in which the sudden extinction of many marine groups occurred.

The period was one of great orogenic activity and marine transgression. As North (1980) has noted, the Early and Middle Devonian epochs saw the widespread conversion of 'oceanic' or eugeosynclinal regimes to 'continental', 'orogenic', or 'island arc' regimes through continental collisions and continent/arc collisions. Amongst the conspicuous changes to occur were the closure of the 'Caledonian' Iapetus Ocean and the migration of continental masses northwards. On the western side of the North American (Laurentian) craton occurred the early phase of the Antler orogeny, brought about by the closure of the marginal ocean between the craton and the Klamath–Sierran arc. On the opposite side of the Baltic–Laurentian craton (also known as North Atlantis or Laurussia) the early phases of closure of the Uralian Ocean were taking place as the Angara continent (Siberia) approached. Elsewhere in the 'Far East' there were orogenic phases on the margins of several of the cratonic masses of China and South-east Asia. There were craton–arc collisions between the Indo–Australian sector of Gondwanaland and the Tasman arc and when the latitudinal ocean between Laurasia and Gondwanaland began to open there was a collision between the South American plate and that which now carries Mexico and Central America. The result was to provide many of the continental masses to the north and south of the 'Pre-Hercynian' or 'Palaeotethys' Ocean with marginal mountains with calc-alkaline magmatic features, granitic plutons included.

In North America the Kaskaskia (transgressive)

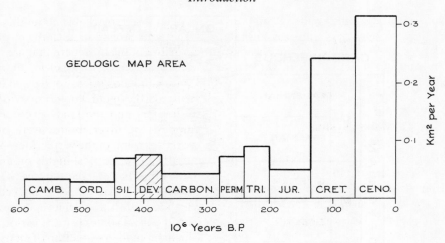

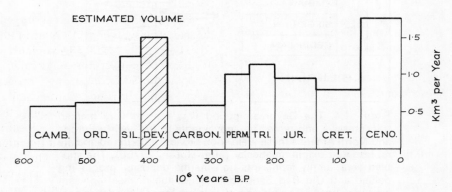

Figure 1.3. Estimated surviving volumes and map areas of the Phanerozoic systems (after Raup, 1976b). Deep-sea deposits are excluded from the estimated sediment volumes. The prominence of the Devonian system among the Palaeozoic is conspicuous. Rates of accumulation per year are estimated on the right and the implication is that subsidence, allowing for the high volume of Devonian sediments, must have been rather more active than during the foregoing and succeeding periods until Cenozoic time.

cratonic sequence reached its zenith, and it can be seen that throughout the Kaskaskia cycle there was more or less continuous orogenic disturbance on both the east and west continental margins of North America. On its northern flank the Franklinian orogeny was in progress so that around almost its entire circumference the continent was being assailed. At no other time was this to occur.

The Stratigraphic Column

Before embarking upon a survey and discussion of Devonian rocks, their origins and the phase of earth

history that they represent, it may be useful to review aspects of some necessary basic concepts. To many geologists stratigraphy is a slightly disreputable member of the family of geological disciplines. It seems to lack that kind of basic order and regularity of features that are so important to physical scientists, but as it is concerned with the description, organisation and classification of stratified rocks the impression is hardly surprising. Much of the continental surface of the earth does indeed show a bewildering array of strata. Among the earliest efforts of geologists were attempts to classify strata on the basis of their apparent composition, their order of succession and their tectonic

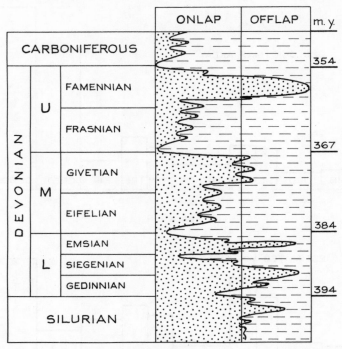

Figure 1.4. The Devonian period was one of very marked transgressive and regressive movements of the sea. In Europe and North America these were soon broadly recognised and are represented diagrammatically (after House, 1975b). The stippled area largely represents non-marine deposits, prograding deltaic and Old Red Sandstone facies; the dashes represent marine deposits. The placing of series or stage boundaries near to or at marked facies changes is common but has led to some confusion where the facies changes occur at different levels from place to place. The dates, given by Gale *et al.* (1979), emphasise the rapidity of the reversals of sea movement in the Early Devonian and near the end of the period.

condition. From the attempts of Lehmann, Hutton and others in the eighteenth century to establish a universal classification and a unifying theory of the origins of rocks came the basic tenets of modern stratigraphy. Hutton, demolishing the Neptunist philosophy, established the vital principles of uniformitarianism and of superposition. In Hutton's day it was no mean achievement to initiate such a radical concept that immediately provided a new means of investigating earth history. There followed from the work of William Smith, carried out at the opening of the eighteenth century, the recognition of stratal units over wide areas on the basis of their included fossils. Palaeontological correlation, the beginning of biostratigraphy, was

demonstrated to be a prime means of matching successions in different regions. Comparative stratigraphic studies could be placed upon a new footing and geological time could be seen as marked by the passage of life from one state to another. Catastrophism was still invoked to explain widespread unconformities.

Lyell achieved a major step when in his *Principles of Geology* in 1833 he published a stratigraphic column which is the recognisable ancestor of the one in use world-wide today (see figure 1.5). He utilised Smith's rock units, added others from Europe and arranged them in a sort of hierarchy, 'groups' and 'periods' etc., covering all of what we now term the Phanerozoic eon.

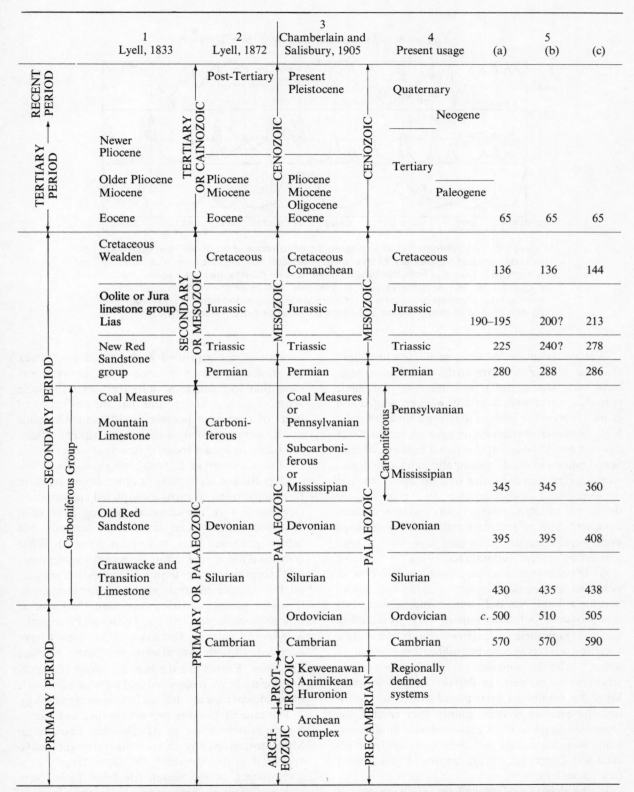

	1 Lyell, 1833	2 Lyell, 1872	3 Chamberlain and Salisbury, 1905	4 Present usage	5 (a)	(b)	(c)
RECENT PERIOD		Post-Tertiary	Present Pleistocene	Quaternary			
TERTIARY PERIOD	Newer Pliocene			Neogene			
	Older Pliocene Miocene	Pliocene Miocene	Pliocene Miocene Oligocene	Tertiary			
				Paleogene			
	Eocene	Eocene	Eocene		65	65	65
SECONDARY PERIOD	Cretaceous Wealden	Cretaceous	Cretaceous Comanchean	Cretaceous	136	136	144
	Oolite or Jura linestone group Lias	Jurassic	Jurassic	Jurassic	190–195	200?	213
	New Red Sandstone group	Triassic	Triassic	Triassic	225	240?	278
		Permian	Permian	Permian	280	288	286
	Coal Measures		Coal Measures or Pennsylvanian	Pennsylvanian			
	Mountain Limestone	Carboni- ferous	Subcarboni- ferous or Mississipian	Mississipian	345	345	360
	Old Red Sandstone	Devonian	Devonian	Devonian	395	395	408
	Grauwacke and Transition Limestone	Silurian	Silurian	Silurian	430	435	438
PRIMARY PERIOD			Ordovician	Ordovician	c. 500	510	505
		Cambrian	Cambrian	Cambrian	570	570	590
			Keweenawan Animikean Huronion	Regionally defined systems			
			Archean complex				

Figure 1.5. The evolution of the stratigraphic column since Lyell's pioneer *Principles of Geology* (1833). Age dates: (5a) U.S.G.S.; (5b) Van Eysinga (1975); (5c) Harland *et al.* (1982). 'Reliable' age dates first became available with the work of Arthur Holmes shortly before the Second World War. Several tables of Phanerozoic dates appeared between then and 1975 as analytical methods improved. The last column represents the work of very many authors in Britain and the USA.

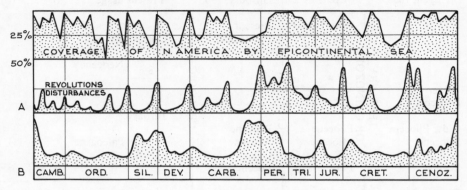

Figure 1.6. (A) Schuchert (1915) plotted the occurrence of orogenic 'revolutions' or disturbances and the amount of the North American continent that was covered by transgressive seas and from this was inclined to view diastrophism as a useful means of correlation. (B) Another version (Umbgrove, 1947) of the variable intensity of diastrophism places more emphasis on more widely separated periods and minimises the occurrence of diastrophism at system boundaries (after Weller, 1960).

Wider correlation led some geologists to believe that the greater or more striking unconformities were very widespread if not universal. Orogenic episodes, world-wide in extent, were envisaged and correlation on the basis of diastrophism achieved a high degree of respectability and acceptance (see figure 1.6). Although the original concept of universal phases of diastrophism affording a means of world correlation has fallen into disfavour, the use of widespread unconformities (see page 36) to define 'natural stratigraphic' units and to document continent-wide or greater inundations and perhaps even world tectonic events (see page 38) has been revived by several authors (Ki, 1975).

As the nineteenth century drew on, the flow of published descriptions of stratigraphic successions became a flood. Not only were monographic treatises produced but short papers were presented in increasing numbers to the growing body of scientific societies, and government publications began to adopt official policies on—and views of—stratigraphic matters. In Britain, Sedgwick established his Cambrian System and Murchison founded the Silurian System; jointly they erected the Devonian System. The Catastrophist school meanwhile was discredited and the biostratigraphic record was continually being improved and refined (see figure 1.6).

In the debate and excitement of those days it became apparent that alternative and rival classifications of the stratified rocks could hold back progress rather than enhance it. Moreover it was clear that specified rock units represented specific spans of geological time and that units of rock and units of time were necessarily different. Orogenic events were regarded as short-lived violent affairs, separated by longer spans of time that were marked by slower transgression, weathering and deposition.

Over the last sixty years or more, however, there has been a trend towards a simplified classification (see figure 1.7); the classical systems have been accorded priority and have been retained but others, proposed later, have been dropped. What remains is merely contrived for general convenience and acceptance rather than with any idea of making all the designated stratigraphic systems equal in terms of geological time, maximum thickness or other non-biological criteria. The terms Palaeozoic, Mesozoic, Cenozoic, Tertiary and Quaternary have survived, though not always with their original meanings. Fossils are the tags by which biostratigraphic units are recognised and by which intervals of geological time are defined (chronostratigraphy).

The mid part of the present century has seen a tremendous increase in stratigraphic knowledge. Magnetostratigraphy and seismostratigraphy have appeared in the literature, the latter largely as a consequence of the search for fossil fuels. New interpretations of orogenic events and the advent of the theory of plate tectonics has profoundly affected

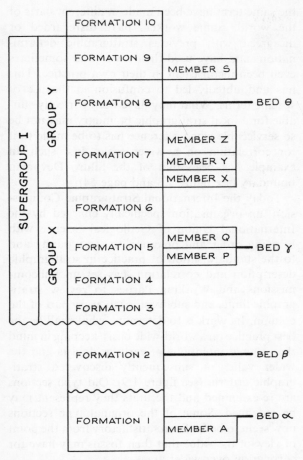

Figure 1.7. Stratigraphic classification remains largely based upon mappable rock units. Lithostratigraphic classification includes the fundamental categories shown here (after Holland, 1978).

treatment of sedimentological data helps in the construction of some palaeoenvironmental models but the field for this is also rather limited.

Computer-aided studies of the stratigraphic development of particular sedimentary basins, using the great volume of data from subsurface exploration, have achieved more detailed views of facies changes, variations in thickness and other properties than was ever possible previously. All of this work has to be based upon the proper application of proved stratigraphic principles and on consistency in the use of stratigraphic criteria (Hedberg, 1961).

From time to time geologists express some frustration at their inability to bring greater precision to stratigraphy either in terms of biostratigraphic correlation or in terms of absolute rock dating. The geological periods are of different durations: the Cambrian period is regarded as having lasted about 100 million years; the Triassic a mere 35 million years. The Palaeozoic era lasted about 375 million years, the Mesozoic 160 million years and the Cainozoic has endured for 65 million years. There seems to be little chance that geologists will ever use a time-scale in which the geological eras and periods were consistently of an equal age, with a 'standard period' of say, some 40 million years, useful though that might be.

Fischer (1979) and others have directed our attention to a variety of cycles that occur in earth history, affecting such features as sea level, climate, ocean and atmospheric composition. These in their turn may reflect cyclic or at least recurring changes in the intensity of mantle convection, but while they are 'natural cycles' they are too large to be of immediate use in stratigraphy.

A concept that has persisted with some vigour has been that of world-wide 'natural breaks' in the record, and this has been particularly espoused by geologists of the U.S.S.R. They have in their stratigraphic philosophy placed an emphasis on 'natural steps' in earth history and on the idea that stratigraphic subdivisions should be based on the totality of *all* lines of evidence, not solely lithostratigraphy or biostratigraphy. Separate kinds of classification, based upon a single criterion or selected criteria, would be to them unacceptable. Yet there is still far from universal agreement that world-wide natural breaks or steps of sufficient frequency to be geochronologically useful can be recognised.

our views of the influence of tectonics upon sedimentation and stratigraphy itself. Most recently 'event stratigraphy'—the calibration of the effects of floods, storms, volcanic eruptions, magnetic reversals, sudden climatic or oceanographic changes, asteroid impacts and of course biological changes and extinctions—has been identified as providing a calendar of the evolution of the continents and oceans.

It could be said with some justification that the discipline has not progressed far from the qualitative stage to the quantitative. The applications of statistical methods to stratigraphy have not been numerous or particularly successful. Statistical

Although it is not surprising that this use of many kinds of data has proved to be an unattainable goal, the International Commission on Stratigraphy has set up a working group on a unified stratigraphic time-scale (UTS). Its aim is to produce a time-scale that combines absolute and relative time-scales—one which has the greatest global applicability. In this the working group will need 'to examine the contribution of stratigraphy, palaeontology, geochronology and fossil magnetic polarity ... and to suggest ways and means by which regional scales can contribute to a world-wide time-scale' (Van Hinte, 1978). So far, progress has been minimal and this ideal appears to be achievable only in the very long term, geologists for the most part being somewhat unwilling to commit their energies to goals so seemingly remote.

Stratigraphic Procedures

It is sometimes not realised to what extent the stratigraphic column is based upon actual rock successions, as measured and described from particular localities. At such *type localities* have been recorded the *type sections* upon which systems, series and other stratigraphic units have been based. Not all geologists have been very careful in defining the limits of the units that they have distinguished in one part of the world or another. Different usages of

the same term have been made in different parts of the world. Some workers have disregarded or disagreed with previous stratigraphic determinations and have used their own—and sometimes even been inconsistent in their own practice. This has undoubtedly led to confusion in the use of specific terms. What may have been perfectly suitable for a local stratigraphic boundary may not be so serviceable when reference has to be made to it for correlation on a wider scale. Just such an example has been that of the Siluro–Devonian boundary (see figure 1.8 and page 54).

Today the International Stratigraphic Commission, an organisation specifically charged by the International Union of Geological Science with bringing order and precision to stratigraphy and to the standardisation of practice in stratigraphic description and correlation, has set up Subcommissions and Working Parties to review stratigraphic limits and subdivisions in every part of the column. Its work is to determine what shall be the best practice on a world-wide basis, keeping in mind the original intentions of early geologists and the wider value of subsequently discovered stratigraphic criteria (see figure 1.9). Old type sections are re-examined and the units they represent may be redefined. Some of the original type sections now seem to be far from satisfactory from the point of view of the value that their fossils may have for correlation over wider areas.

Figure 1.8. The type section at Klonk in Czechoslovakia where the boundary between Silurian and Devonian systems is by agreement drawn at the base of bed 20—the site of a 'golden spike' that defines the base of the Devonian system. Black—greyish-black calcareous shales; 'bricked'—greyish-black and dark-grey, fine-grained platy micritic limestones; dotted—paler grey, granular fine-grained limestones; 'jackstrawed'—coarse detrital crinoidal limestone with some turbidite component (after Chlupáč *et al.*, 1972).

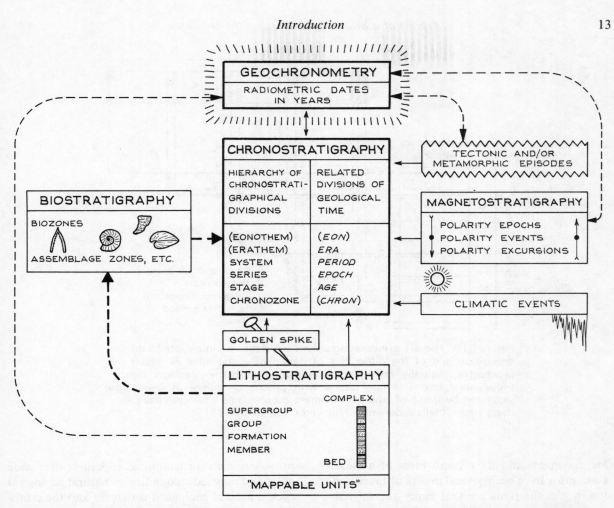

Figure 1.9. The procedures of stratigraphy and the categories within it as indicated by Holland (1978). The chronostratigraphical record is the central repository for the data derived from the procedures and phenomena indicated around them. The chronostratigraphic terms shown in brackets are rarely used but, logically, could be used to a greater extent.

Breaks in the Succession

It was soon apparent to the pioneers in geology that the succession of strata at any place may be incomplete either because no deposition may have occurred over part of the time or because rocks already deposited may have been removed by erosion (see figure 1.10). Such breaks can vary in size from very small to those that involve a major span of geological time.

The smallest of all breaks in stratigraphic successions — Barrell (1917) in the U.S.A. first referred to them as *diastems* — are undoubtedly very numerous, and the majority are undetectable. Bedding planes must in the majority of cases mark diastems, and show interruptions or changes in deposition. Thinly bedded formations may then represent intermittently deposited sediments with slight but repeated changes to the supply of sediment, or to the transporting agency or in the depositional surface that occurs time and time again.

Barrell calculated that the mean rate of sedimentation throughout Phanerozoic time has been very slow, about 0.2 m per 1000 years. The systems, however, rarely show a thickness that approaches

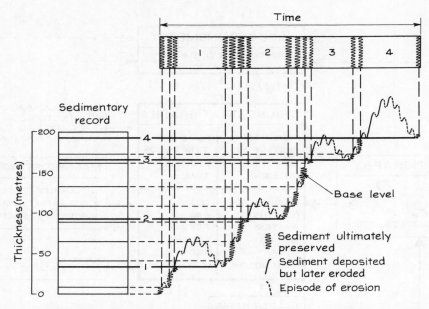

Figure 1.10. There is an increasing awareness of the very interrupted nature of deposition and of the difficulties of determining the rates at which the process(es) actually take place. Deposition in many—perhaps most—sedimentary basins is intermittent with phases or pulses of deposition separated by those of erosion. Sediment accumulates when the base level rises; when it falls sediment is removed (after Barrell, 1917).

this maximum and their mean rates of accumulation must have been several orders of magnitude slower. Yet the signs are that many ancient sediments accumulated rapidly, just as do their modern counterparts. Between the deposition of individual beds of sediment there may have been phases of erosion or non-deposition, so that apparently continuous sedimentation was in fact episodic — a succession of separated 'pulses' or quanta.

Unconformities are the most striking natural breaks in the local stratigraphic record and some unconformities are traceable over enormous areas (see figure 1.11). Followed from one region to another, concordant breaks or *disconformities* may cut across biostratigraphically or otherwise verified isochronous horizons and so cannot be true time planes. Between such major widespread unconformities the successions approximate in scale generally to supergroups and span more than one geological system. Ki (1975) proposed the formal term *synthem* for such unconformity-bounded bodies of strata and suggested that they may

provide a basis for tectonic correlations over wide areas and may aid recognition of natural geological cycles, natural geological provinces and the establishment of a natural stratigraphic classification.

In the case of unconformities it is usual for a significant amount of erosion to have occurred as well as non-deposition of sediment, especially where an *angular unconformity* occurs. How long each of these processes may have operated is commonly difficult to determine. Among the first unconformities to be described in scientific terms was that of the Upper Old Red Sandstone on deformed and eroded Silurian rocks at Jedburgh in the Scottish borderland. The Jedburgh example shows a more or less flat plane of unconformity and such planar surfaces are often associated with marine erosion (though not in fact in this instance). Marine sediments above the unconformity would add weight to that argument. However, highly irregular surfaces, sometimes known as *buried landscapes*, are found elsewhere and these are especially typical below continental formations. The unconformities be-

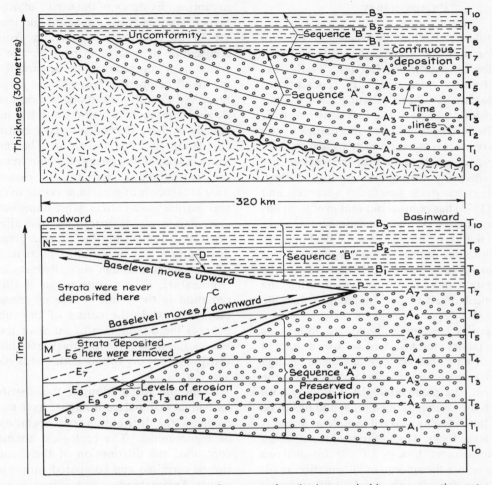

Figure 1.11. The origin and history of an unconformity is revealed in cross-sections at the edge of a sedimentary basin. The sequence 'A' rests upon eroded igneous rocks and is overlain unconformably by sequence 'B' as a result of movement at the basin margin. The analysis of the unconformity between the two sequences with time rather than thicknesses is shown on the vertical scale. As Eicher comments, point C represents a locality where base level intersects the earth's surface so that to the left erosion takes place and to the right deposition may occur: 'At time T_6, when the base level–surface intersect is at point C, erosion has removed the record to the level of dashed line E_6.' At time T_7, the process reverses and deposition is progressively re-established to the left of point P. Deposition spreads over the surface NP which is also the erosional surface LP, the surface of the unconformity (after Eicher, 1976, modified from Wheeler, 1964).

neath the Middle Old Red Sandstone of Scotland or the Lower Old Red Sandstone of south-west Wales are good examples of this.

Thus where the sea is advancing its limits and progressively submerging the land, a *marine transgression* takes place. Where, on the other hand, the land is rising, a *regression* will occur.

At this point it must be said that transgressions and regressions can be indicated by simple lateral or vertical lithological change. The upward change towards deeper-water sediments in a succession or from terrestrial to marine will indicate a transgression; regression will be indicated by trends that are the reverse of these.

Palaeontology

The scientific description of fossils has been in progress for over 200 years and the flow of new palaeontological data in the last few decades has been continually increasing. In consequence palaeontologists have been debating anew a number of questions that lie at the very centre of their subject. The foremost concerns the nature of the evolutionary process as it is revealed by fossils: there has also been argument about the apparent increase in species diversity throughout geological time. Palaeoecology has made significant advances, and Valentine (1973) has pointed out that evolution has interactive biological and environmental components so that it is therefore an ecological process. The palaeontology of the Devonian system records a particular stage in this process and reflects the environmental conditions of the time as well as the morphological diversity and abundance that life had achieved.

In comparing ancient and modern faunas or floras the palaeontologist has to use a limited collection of data, but ecology has made enough progress for many of its precepts to be applied to the fossil record. The palaeontologist is usually concerned with the neritic marine fauna, and has to be content with the highly selected sample of it that is available in the rocks. The vagaries of fossilisation and the subsequent history of the fossiliferous formations reduce the amount of information available. We have, for example, a relatively extensive knowledge of the brachiopod faunas of the Devonian tropical and temperate realms—less of the cool-water (southern hemisphere) Malvinokaffric realm—but little knowledge of the soft-bodied marine organisms of that time in the tropical or other areas (see figure 1.12). While we may understand the form and function of many fossils from the Devonian, many of the influences that moulded them will remain unknown.

Amongst Palaeozoic systems the Devonian is remarkable for the taxonomic diversity of its fossils: the full extent of this diversity is a matter of debate but our estimate of it is influenced by what we know of the fossils of earlier systems and by our knowledge of Recent biosphere. Around the continents of the tropical and temperate latitudes the Devonian seas were densely populated by a wide range of organisms. Evidence of the neritic marine fauna of higher latitudes is less abundant and the nature of life in the Devonian oceans far from land is conjectural to a large degree. Neritic benthonic communities ranged from the apparently simple to the highly complex so far as we can judge from the preserved (largely calcareous skeletonised) forms. Pelagic communities would doubtless have varied equally widely, dependent upon latitude, water circulation, aeration and so on. The currents and water masses of the oceans provide a framework for the distribution of environmental parameters and their influence is of great importance to continental and marine biotas alike. Most immediately, however, it is ocean temperatures that affect marine organisms, influencing their forms, functions and distribution. Ocean salinity, nutrient salts and oxygen content, though normally less variable than temperature, have a wide range of effects. Solar radiation is the prime source of energy for the biosphere, while the nature of the substrate influences the distribution of much of the benthos. Waves, tides and currents influence the life-style and distribution of organisms both benthonic and pelagic.

From the recognition of the distributions of individual taxa palaeontologists have moved on to recognise the distributions of associations of taxa, or communities. The ecological conditions that controlled the distribution of these communities can be surmised and compared with modern analogues. Ancient biogeographical provinces—regions in which the communities maintain a characteristic composition—can be suggested from surveys of community distribution. For some of the best known and most intensively studied fossil groups in the Devonian system biogeographical provinces have been suggested. The degree of correspondence between the provinces of different Devonian groups such as brachiopods and corals is gratifyingly as close as might be expected (see Boucot *et al.*, 1968; Oliver, 1976).

During the 50 million years or so of the Devonian period great changes in physical geography profoundly altered biological distributions. Biogeographical provinces changed in area and shape and, insofar as can be judged from the marine neritic benthos, there was a general decrease in provinciality amongst the marine faunas. There was a

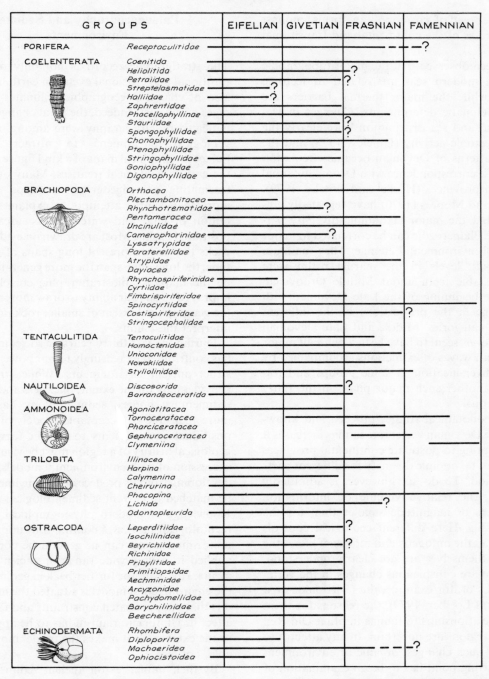

Figure 1.12. The Devonian marine faunal record includes many extinctions among the common taxonomic groups, especially at or near the end of the Givetian stage (after House, 1975a). Even allowing for the very large number of marine forms within the system (see page 65) these extinctions are highly significant and call for a special explanation (page 187).

coincident spread of 'cosmopolitan' forms amongst the vertebrates of the continental and transitional realms.

It has been observed that the major topographic barriers in modern seas are related to tectonic processes while the major thermal barriers are related to climatic determinants. Changes in Devonian land and sea distribution were clearly the result of tectonic activity (Dineley, 1979) and the possible patterns of Devonian oceanic circulation show a close correspondence with Devonian faunal realms and provinces (Heckel and Witzke, 1979). Valentine and Moores (1970) have repeatedly emphasised that the major taxonomic diversity patterns of the Phanerozoic can be correlated to major patterns of environmental change. They maintain that diversity levels in the marine realm were relatively stable from about Middle Ordovician time to the beginning of the Late Devonian with little change in the proportions of the different taxonomic categories. Middle and Late Devonian times, however, seem to have been a phase of major changes with waves of extinctions accompanied by vigorous diversification in some groups and followed by another such major phase in the Early Carboniferous.

As palaeobotanical studies proceed and knowledge of the Devonian vertebrate faunas increases it is also tempting to postulate continental provinces and levels of taxonomic diversity for different parts of the period. To do so, however, is admittedly hazardous, the relative volume of information being orders of magnitude smaller than for the marine fossils. Here different ecological controls operated and the ontogeny and modes of life of the organisms themselves are not clearly understood. While there are conspicuous changes in the vertebrate faunas of, for example, the Caledonide area (Dineley and Loeffler, 1979), the reasons for them and their relationships to faunas in other Old Red Sandstone regions are uncertain. In any attempt to understand such changes, organic or environmental, our starting point has to be an examination of the rocks themselves, the fossil-enclosing sediments and their relationships.

Palaeogeography and Sedimentary Environments

The stratigraphic record is not only one of depositional and erosional events in earth history but also of changing geographies, climates and ecologies of the past. Indeed, the great changes that are indicated by stratigraphy were among the earliest geological phenomena to attract attention. Palaeogeographical maps of a kind figure in some of the earliest geological treatises. Many of the most fascinating palaeogeographical maps are much more than simple attempts to explain the distribution of land and water, source area and depositional basins. Most are drawn on evidence from rocks that may represent long spans of geological time; the longer the span the more generalised has to be the map. As biostratigraphic correlation has improved so has our ability to draw more accurately the geographical extent of smaller rock units and to interpret these units.

During the last thirty years or so sedimentology has contributed increasingly to our power to reconstruct ancient environments. While stratigraphy provides data on the extent, thickness and nature of sedimentary bodies, sedimentology has enabled more plausible or acceptable models of their depositional environments to be put forward. The geological history of a region must be visualised as a succession of such environmental models, each one developing from its predecessor along lines that can be matched in current sedimentological processes. Knowledge of modern physiographical and sedimentological processes continues to improve, and uniformitarianism can in general be convincingly applied to a very wide range of ancient environments. However, the further back in geological time the environmental model is situated the more likely it is that unappreciated constraints and factors will affect it. Uniformitarianism has to be applied only to the extent that it makes the model more acceptable, not less.

By their influences on ancient sedimentary environments, palaeoclimates have left many marks upon the record and our understanding of the

Quaternary or Cainozoic record is better than it can ever be for the Palaeozoic. Climatic changes and their causes are of direct concern to geological history. Modern studies of deep-sea sediments reveal a continuous story of climatic change that is so far not paralleled by any such Palaeozoic record. It has, nevertheless, to be said that Cainozoic analogues may not be apparent for Palaeozoic climates. Beyond the terrestrial record of geological processes there may be questions of extraterrestrial influences. Has the solar constant been stable throughout the Phanerozoic? Has the sun's radiation varied, even oscillated, regularly with time? Have other extraterrestrial events been influential in earth's evolution? Clearly the sky is not the limit in palaeogeography.

2

The Materials of Stratigraphy

Historical geology involves concepts of time, life and environments that are unfamiliar to us in terms of personal experience, but it is founded upon the study of materials and of structures that accumulated unevenly across the earth and at greatly varying rates in time. The stratigraphic column provides these basic materials, not only as sedimentary rocks and fossils but also as igneous and metamorphosed rocks of many kinds. Studies of clastic, carbonate, volcanic and metamorphic rocks have become so specialised that the significance of events that led to the accumulation of a varied pile of rock units may be missed by a specialist who is working in isolation. Help from the structural geologist, the palaeontologist or the geochemist may be needed to explain the evolution of a succession of rocks. Nevertheless, the unravelling of the record begins with the study of its sedimentary or volcanic deposits. Investigation starts with lithological and stratigraphical (or structural) features, proceeds to sedimentology and palaeontology, perhaps to experimental and comparative studies, and hopefully reaches a point where hypotheses can be put up like 'Aunt Sallys'. Hypotheses should be bold affairs that not only account for the facts but also stimulate further interest in the topic. They can be effectively knocked down only when there is something better to replace them.

Only around 5 per cent of the known lithosphere (15–20 km thick) appears to consist of sedimentary and metasedimentary rocks, but some 66 per cent of the continental areas is covered by them. The thickness of these sedimentary layers ranges up to almost 15 km with an average of rather more than 1.5 km over the continents. Calculations of the total mass of sediments and sedimentary rocks have been between $(4.1 \pm 0.6) \times 10^8 \, \text{km}^3$ and $15 \times 10^8 \, \text{km}^3$. The latter is probably nearer the truth. Pettijohn (1975) suggested that the relative proportions of the three common sedimentary types are shales 58 per cent, sandstones 22 per cent and carbonates 20 per cent. Krynine (1942) thought that about 30 per cent of the 'average sediment' is derived from the reworking of older deposits, but Blatt and Jones (1975) estimated that 80 per cent of the grains in sandstones are derived from older sedimentary rocks.

The fragmental or clastic rocks can be analysed in terms of their component grains, matrix and cement, and a time-independent model for their formation can be set up. The carbonate rocks, however, are largely the result of organic activity, a process that has evolved with time (see figure 2.1). During Cambro–Ordovician, Devonian and Cretaceous times the deposition of exceptionally widespread quantities of carbonate sediments occurred. Chemical and physicochemical activity has been important in the production of evaporites, glauconitic, phosphorite and ferruginous beds. Determining how continuous those processes were and for how long they endured is, however, difficult. Clastic and carbonate rocks both require the expenditure of energy for their formation and they may be deposited in adjacent parts of the same basin, their contact being abrupt or gradational. Sedimentary contacts denote environmental boundaries or changes with time, depending upon whether the different lithologies are in horizontal (isochronous) contact or vertical (consequent) contact. Thus sedimentary environments are ephemeral or temporary phenomena, changing many of their characteristics throughout geological time. We must allow that many ancient environments were different from those of today and we are also puzzled to find particular kinds of sedimentary

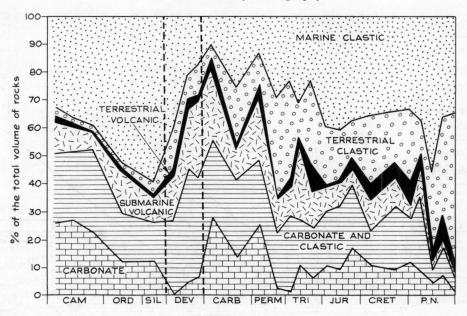

Figure 2.1. Ronov *et al.* (1980) made an estimate of the changes in the ratio of the lithologic associations within the Phanerozoic rock succession which indicates an exceptional reduction in the proportion of marine clastics during the Devonian and an all-time low in carbonate accumulation early in Devonian time. These are interesting in the light of evidence for a progressive transgression over the continental areas during the period.

rocks that are seemingly of about the same age wherever they are found. Their genesis may have depended upon time as well as space, and upon the biosphere as well as physical processes.

Volcanic accumulations play a significant and special part in the geological record, indicating the kind of crustal unrest that is present in an area and the proximity or location of actual vents or fissures (see figure 2.1). Even so, the volcanological and tectonic information that such rocks provide is of a more restricted kind than that offered from sedimentary formations. Bentonites and tuff-falls prove of value in correlation.

Plate-tectonic studies imply that at times 'supercontinents' were a major influence upon climate, ocean circulation, provinciality and other biological features; at times orogeny was exceptionally widespread and geography was changing rapidly, and at times continental dispersion influenced the evolution of sedimentary rocks. Are patterns and long-continued trends in sediment production discernible and has sediment production diversified with

the passage of time? To respond to these questions is difficult and the answers perhaps are not very meaningful. To be concerned with the 'sedimentary palaeogeography' of a single period is only a slightly easier commitment. The answers to questions about it are not always clear or unequivocal.

The migratory nature of many depositional environments is well illustrated by marine sedimentary rocks. With rise or fall in sea level the areas in which these contemporaneous sediments accumulate migrate. Transgressions and regressions occur; sediments that were deposited originally side by side may then also occur in vertical sequence, as stated in Walther's Law (see Middleton, V. W., 1973) (figure 2.2). Where such sequences occur in stratigraphy it is common for the geologist to have difficulty in establishing good biostratigraphic correlation because the different depositional environments probably supported quite different assemblages of organisms. To construct the sedimentary history of a depositional area good palaeontological correlation, as independent of lithological con-

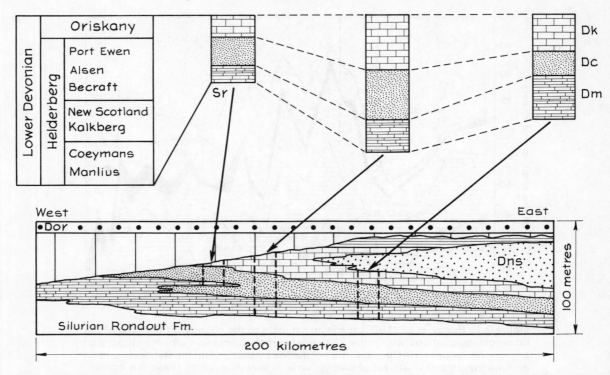

Figure 2.2. Walther's Law states that in transgressive successions the vertical order of facies is also that of facies in lateral aspect. Laporte (1969) pointed out a good example of this in the Helderberg Group of marine limestones and shales in New York State. The distinctive fossils of each facies reflect different environments of deposition and not, as was once thought, different spans of Devonian time. Dm, Dc, Dk and Dns are the Manlius, Coeymans, Kalkberg, and New Scotland formations respectively. Dor is the Devonian Oriskany Sandstone which rests unconformably on the Helderberg. Sr is the Silurian rondont formation.

trol as possible, is essential. Conversely, to understand the environment in which lived the organisms that became fossils the palaeontologist needs to study sedimentology.

Habitats are not static. There is a flow of energy through each one that influences its physical character and its biological content. The level and nature of that flow of energy may be indicated by the sediments that accumulate, or by gaps between successive sediments. There may also have been a vigorous biological productivity that is scarcely indicated by the sediments and their fossil contents.

Sedimentation Rates and the Stratigraphic Record

In trying to understand ancient environments in

terms of their modern analogues the comparison of Recent sedimentation rates with the thicknesses in the stratigraphic record may seem to offer little help. Rates of sedimentation are clearly affected by tectonism, weathering and transport and by biological activity. The physical and chemical processes of rock weathering and sediment transport are thought to have been little different in Early Palaeozoic times from what they are today. Certainly the laws of physics are likely to have been the same. On the other hand, as the diversity of organisms and their activities have increased, the biological processes that lead to sediment formation may now be more varied and effective than they were. The vegetational cover of the continents probably first became extensive during mid-Palaeozoic time and local rates of erosion may have dropped accordingly (see Gregor, 1970). Surveys of the extent, existing or original volume and kinds of rocks that are formed over any particular interval of geological time are still only approximations. Undeterred, several authors have attempted this task in the hope that new evidence will come to light concerning the rate of geological processes or the influence of this or that factor (see page 24).

The global rate of deposition of sediments today is enormously difficult to assess; clearly different rates prevail in different sedimentary environments. Contemporaneous subaerial erosion is a process that can be directly discussed but its common counterpart, subaquatic sedimentation, is less easily studied (see figure 2.3). Much, perhaps most, *normal* sedimentation is a slow discontinuous process with sediment accumulating between phases of non-deposition and even erosion. Many believe it occurs dominantly as small quanta or pulses in most environments of clastic deposition. In some continental semi-arid or glacial environments these pulses may be annual floods or changes of water level. In many environments, however, it is seen as a genuinely continuous process. Many pelagic carbonate successions, for instance, are demonstrably continuous.

At the other extreme, *catastrophic* sedimentation occurs exceptionally and involves sudden activity on a grand scale. Turbidites laid down on abyssal plains occur through this kind of event (see page 157), and between successive turbidites there may appear to be a hiatus in deposition—a hiatus of unknown length.

Ager (1973) concluded, when discussing his proposition that 'the sedimentary pile at any one place on the earth's surface is nothing more than a tiny and fragmentary record of vast periods of earth history', that the phenomenon of the gap is more important (that is, it may be of greater time significance) than the record. This is, however, not to say that local estimates of rate for some kinds of sedimentation cannot be found. Deltaic environments are commonly thought of as receiving detritus at an immense rate, while pelagic (oceanic) environments receive it at an extremely slow rate. Ager's examples when discussing 'catastrophic stratigraphy' are again pertinent, and he emphasises

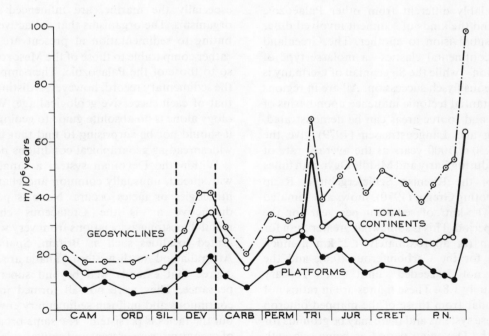

Figure 2.3. Ronov *et al.* (1980) estimate the variation with time of the average rate of accumulation of sediments within 'geosynclines', platforms and continents overall. High rates are found in the Devonian, Permian, Triassic and Cainozoic (P.N.) where they are said to be peculiar to the initial and final stages of tectonic cycles.

that sedimentation in the past has often been very rapid indeed and very spasmodic.

In estimates of the surviving areas and volumes of sedimentary rocks (Blatt and Jones, 1975; Raup, 1976b; Ronov *et al.*, 1980) the Devonian has relatively high values (see figure 1.3). Gregor (1970) suggests 'that the mass–age distribution of Phanerozoic sediments . . . can be expressed in terms of survival rate (surviving mass of each System divided by the duration of the corresponding Period) plotted against median age. Such a plot shows a general decline of survival rate with increasing age.' For post-Devonian systems the decline is nearly exponential.

Friend and House (1964) gave the maximum known thicknesses of Devonian subdivisions as follows:

Famennian	Arctic Canada	10000 ft (3330 m)
Frasnian	New York State	6000 ft (1812 m)
Middle Devonian	Greenland	13000 ft (3933 m)
Emsian	Germany	7000 ft (2120 m)
Siegennian	Germany	20000+ ft (6660 m)
Gedinnian	Ardennes	5500 ft (1600 m)

Assuming a value of 50 million years for the period, these indicate rates of deposition that are not appreciably different from other Palaeozoic instances and the kinds of sediment involved differ from one subdivision to another. The Greenland pile is of continental clastics—a molasse type of accumulation—while the Siegennian of Germany is an argillaceous flysch succession. All are in regions where substantial tectonic influence upon basins of deposition and source areas can be demonstrated.

Goldring and Langenstrassen (1979) give the figure of 0.30 m/1000 years as the average rate of deposition during Early and Middle Devonian times in parts of the Rhenish Schiefergebirge. Raup (1976b), quoting Gregor (1970), shows an estimated volume of 1.5 km^3 of sediment per year for the Devonian period. This is somewhat greater than for the Silurian (at approximately 1.25 km^3), much more than for the Carboniferous (0.6), and the value is not exceeded until the Tertiary (approximately 1.8). These figures are in ratios not very dissimilar from those of the mapped outcrop areas of these systems and there may be grounds for regarding the Devonian period as being one with unprecedentedly high sedimentation rates. This is perhaps to be expected in a system that indicates world-wide transgressions, extensive and prolonged orogenies and volcanic activity and a geographical configuration of land and sea that would lead to regions of high precipitation in warm to temperate latitudes. High rates of weathering would produce in them large quantities of regolith and clastic material for transport to the seas. The Old Red Sandstone is at least in part witness to this. In addition to clastic sedimentation on a grand scale, carbonate deposition by organic activity in shallow, warm, shelf seas was rapid and widespread (see Heckel and Witzke, 1979).

Devonian Rocks

As comparative stratigraphy proceeds, it is increasingly a matter of comment that rock types, especially the sedimentary, vary with age. Ager (1973) has written on the 'persistence of facies' in the Upper Cretaceous, the Jurassic and the Palaeozoic systems. He concluded that 'at certain times in earth history, particular types of sedimentary environment were prevalent over vast areas of the earth's surface'. Many Recent sedimentary environments, especially the marine, are influenced by living organisms. The organisms that are actively contributing to sedimentation at present are, in total, rather comparable to those of the Mesozoic but less so to those of the Palaeozoic. Their imprint upon the sedimentary record, however, is distinctive, as is that of each successive geological age. While lithology alone is no absolute guide to geological date, it should not be surprising to find that imprint as widespread as geographical conditions permitted.

Within the Devonian system a range of such widespread, unusually common and characteristic lithologies or facies occurs. None is perhaps so distinctive as is the Cretaceous chalk, but reefal–limestone associations from very widely separated localities such as Britain, Spain, eastern Australia and north-western Canada are alike to a degree that extends well beyond superficial appearance. They were not all formed in a single continuous and uniform sedimentary environment but the facies is persistent. The same broad groups of organisms were present and active in similar ways wherever conditions suited, and many of the species were essentially Devonian. Numerous other examples exist.

Mention should also be made of the remarkable continuity and extent of formation character or facies that exists in some regions. The black-shale facies of the Middle Devonian of North America covers thousands of square kilometres and the origin of this euxinic deposit over so great an area poses problems to which interesting responses have been made (see page 121).

Non-carbonate but organic Devonian sediments include those rich in siliceous, phosphatic and hydrocarbon materials. *Cherts* and silica-rich sedimentary rocks occur in many Devonian rock assemblages. Some appear to be largely chemical in origin but the contemporaneous or post-depositional precipitation of the silica is arguable. The distinction made between *cratonic cherts* and *geosynclinal cherts* is important. The former are essentially the deposits of silica in shallow-water environments, the latter are perhaps the ancient analogues of modern deep-sea radiolarian oozes. The Woodford chert (Oklahoma), the Caballos Novaculite of Texas and the Arkansas Novaculite have been studied on account of their broad extent and lithological characteristics. Novaculite is a very dense, even-textured, light-coloured, cryptocrystalline silica rock in which microcrystalline quartz dominates over chalcedony. Essentially it is a bedded chert. Folk (1973) found evidence to suggest that the Caballos Novaculite was deposited in shallow lagoonal environments with abundant radiolaria. Associated with these rocks are giant fossil logs and red jaspers that perhaps represent silicified palaeosols. Folk and McBride (1976), however, offered alternative views on the origin of this novaculite: McBride favoured a deep-water environment while Folk maintained his previous view.

Most of the commercially important phosphatic sediments or phosphorites are associated with marine sediments, especially black shales, which by comparison with modern phosphorite sediments are thought to originate where deep phosphate-rich waters rise towards the surface of the oceans. These deposits are said to mark the eastern sides of ocean floors that were adjacent to lands in the dry tropical belt. Local phosphatic deposits may originate from a secondary enrichment, being residual and associated with solution and redeposition in the vicinity of erosion surfaces. Lag deposits can include much nodular and secondary phosphatic material in limestones, others may be associated with glauconitic sands. All seem to originate in shallow waters. The black Devonian phosphates of Tennessee are thought to have been reworked from residual Ordovician phosphates in the same area. In other instances it is thought that very slow sedimentation and a higher than normal pH will provide the right conditions for phosphate precipitation once an influx of fresh sea water enters the basin.

Bone-beds are commonly highly phosphatic because of the amounts of nodular phosphate material as well as bone present. The famous but minor Ludlow Bone Bed (Antia, 1980) is in effect a lag-deposited thelodont scale sand with nodules and the Middle Devonian bone-beds of the Cincinnati Arch region (Wells, 1944) are also reworked concentrates. Catastrophic annihilations of fish have produced very localised layers of phosphatic vertebrate material in some mid-Palaeozoic formations but they are not stratigraphically significant nor very widespread. North (1979, 1980) distinguishes only four intervals during the Phanerozoic in which effective petroleum-source sediments were deposited widely and in quantity enough to give rise to oil. The earliest of these he names as the Frasnian–Famennian (360–340 million years ago). He finds support for the idea that while 'oil and gas have no doubt always been generated in some quantities, somewhere on Earth, a very rich source sediment is almost as uncommon a feature of the stratigraphic column as a rich oil shale, or a whole formation composed largely of diatomite, radiolarite, or phosphorite'. Source sediments he sees as the deposits of latitudinal oceans that are flanked to north and south by tectonically active continents. They occur as phenomena of an interorogenic but turbulent episode, coincident, in North's opinion, with a cycle of regression–transgression–regression. Maximal deposition of these organic-rich sediments corresponded directly with overall sedimentation rates and with minimum depths of the oxygen-deficient layer. These maxima may also coincide with phases of widespread warm climate and faunal diversification, cosmopolitanism, or extinction (see figure 2.4).

There are some interesting features in the distribution of oil-shale maxima according to North (1980) — they correspond to the low points in the

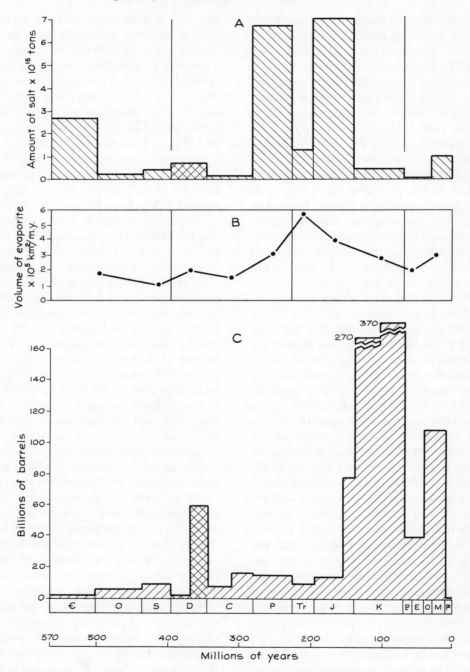

Figure 2.4. Total halite (A, after Kalinko, 1974), all evaporites (B, after Gordon, 1975) and effective oil-source sediments (C), as set out by North (1980). The high correlation between the Devonian entries is conspicuous, and reflects the effective combination of environments, sedimentary formations and subsequent tectonism that occurred in the Devonian period and immediately following.

occurrence of petroleum-source sediments. The latter also correlate with the lowest incidences of marine phosphorite deposition—a surprise in view of the similarity in general conditions that phosphorites and source sediments require. A similar situation appears to exist for the incidence of evaporites.

This work seems to show convincingly that the Late Devonian was an episode of abnormally high petroleum-source sediment deposition, followed immediately by a phase of crustal activity that ensured maturation and preservation of great quantities of oil and gas.

Late Devonian coals are known from a number of localities such as Bear Island in the North Atlantic, the Canadian Arctic Islands, the Catskill mountains of New York State and in central Asia. Macroclastic seams up to 50 cm thick are present in the Canadian Arctic (Embry and Klovan, 1976); those on Bear Island have been worked for some years (Gjelberg, 1977). In strata as old as the Dittonian of South Wales clasts of carbonaceous material more than 1 cm in diameter suggest that microclastic macerals were sufficiently common locally to accumulate in layers that are several centimetres thick. In the Late Devonian of New York State and in China fresh-water swamp vegetation included trees that are more than 10 m high. Vascular plants together with algal and fungal members of the flora formed the bulk of the earliest 'peats'. They seem to have grown rapidly and to have consisted largely of cellulose and waxy substances. Woody tissues first became abundant in Late Devonian time.

Sedimentary Facies and Environments

The term *facies* has been used in a variety of ways since it was first coined by Gressly (1838) (see Krumbein and Sloss, 1963; Middleton, G. V., 1978; Reading, 1978; Walker, R. G., 1979).

Facies relationships are of prime importance in indicating the lateral (geographical) and sequential disposition of the environments that produced them. Breaks in the succession may mark any number of events during which deposition was subordinate to a process of erosion. Reading's (1978, pp. 9–14) admirable discussion of facies lists the prime factors that control the distribution and nature of sedimentary facies as: sedimentary processes, sediment supply, climate, tectonics, sea-level changes, biological activity, water chemistry and volcanism. Of these water chemistry remains the factor that is least obvious in the record of the Devonian period, despite the vast Devonian marine evaporites. Although water chemistry is held to be an important control in lacustrine deposition, the extensive lake deposits are also difficult to interpret.

Cycles, *cyclothems* or *rhythms* have been postulated as relatively common phenomena in many different facies or sequences (see figure 2.5) but the apparent usefulness of the concept of cyclic sedimentation may have been overstressed. The regular and repeated uplift of a source area, subsidence of a basin or changes in climate, sea level or sediment supply are probably very rare. Facies associations (that is, groups of facies that occur together and are genetically related) are commonly easier to treat and to understand than the individual facies alone. The vertical succession of facies in an association may be random or may follow a preferred pattern. A *facies sequence* is where facies pass vertically into one another. It may be bounded at top and bottom by an erosion plane or other signs of a hiatus. Two kinds of sequence are common in clastic formations—coarsening upwards or fining upwards, reflecting increase and decrease in flow-power respectively. Frequently they are obvious to the field worker but it may be advantageous to determine whether or not a particular facies tends to pass into another more often than one would expect in a random arrangement. Reading (1978) remarks that this may not only 'help to detect and define a cyclic arrangement of facies but it may also bring out genetic relationships between facies which might otherwise have been missed' (see also Read, 1973, on cycles in the Pillara Formation, Devonian, of Western Australia).

There are various ways of recording such sequences and of treating the statistical data acquired. Good examples of these techniques, particularly Markov Chain analysis, applied to Devonian alluvial successions are given by Miall (1973) and by Cant and Walker (1976).

Classifications are constructed for different reasons; all are artificial and to an extent designed to emphasise a dominant aspect of the subject. Cur-

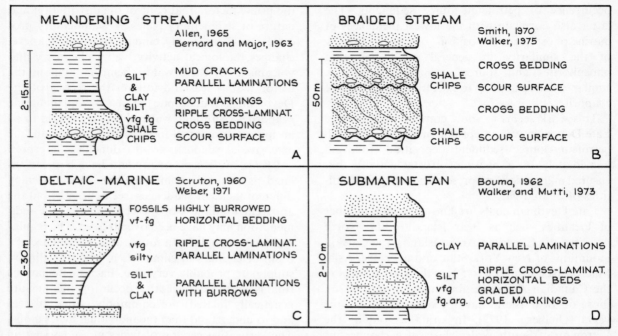

Figure 2.5. Idealised types of lithological cycles within different sedimentary environments (after Embry and Klovan, 1976).

rent classifications of sedimentary environments tend to use a mixture of locality and medium or a process of deposition to establish different categories. A good discussion of the problems of classification is given by Crosby (1972) and Laporte (1979) has surveyed the relationships of many ancient environments, giving some emphasis to palaeoecology and palaeobiological influences.

From Devonian rocks attempts have been made to interpret evidence for sedimentary environments from terrestrial to deep marine, but the system includes many formations that are difficult to understand by reference to modern analogues. The great spread of epeiric carbonates and black shales in the Devonian of North America, for example (see figure 2.6), may be explained by reference to models of ocean circulation that cannot be verified from modern environments. In studies of Devonian palaeogeography it is not always easy to see how a particular environment is related to those that surrounded it or precede or follow it. Questions of spatial and temporal relationships are vital, but many cannot be answered from the stratigraphic

evidence available. The following are the broad categories of sedimentary environments (Reading, 1978):

Alluvial environments	Shallow siliciclastic seas
Deserts	Shallow carbonate environments
Deltas	Pelagic environments
Clastic shorelines	Deep clastic seas
Arid shorelines	Glacial environments

This classification includes a number of discrete deposition models that are both clearly recognisable in modern environments and also inferred from the geological record. Ancient alluvial formations of Devonian age are widespread and well known in the North Atlantic area as the Old Red Sandstone. They also occur in Siberia, China, Australia and Antarctica. Deserts are poorly represented in the Devonian system but minor desert deposits occur in the Old Red Sandstone facies. There are representative Devonian formations from each of the other sedimentary environment categories in vir-

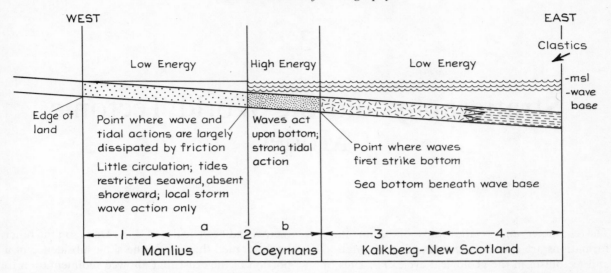

Figure 2.6. Facies analysis of three well-known Devonian formations in New York State suggests the different contemporaneous depositional environments shown. These sediments were deposited in water that was only a few tens of metres deep at most: the horizontal distance however is over 100 km (after Laporte, 1969).

tually all of today's continents. Devonian clastic shoreline deposits, as the heralds of transgressive seas, are known from all continents and arid shorelines are spectacularly represented in the Devonian of western Canada and Western Australia. Shallow carbonate environments (see Heckel and Witzke, 1979) were particularly widespread during Middle Devonian and Frasnian times. Reefs were a conspicuous feature of many of these environments between the palaeolatitudes of 35°N and 35°S.

Pelagic environments are 'open sea', oceanic environments, and pelagic sediments include at least 30 per cent of organic material. It should not be thought that 'pelagic' is synonymous with deep sea. In the stratigraphic record such deposits are commonly associated with ophiolitic rocks but may also extend in thin sheets across the cratons. Deep-sea clastic deposits are essentially resedimented bodies. They include many of the greywacke rocks that are characteristic of geosynclinal regions and as such are widely distributed in Appalachian North

America, Eurasia and eastern Australia. Volcani-clastic deposits may be closely associated with them in great quantity. Several world-wide glacial periods have occurred since Proterozoic times and polar ice may have been present during Devonian. Glaciogenic(?) deposits are reported from South America where they indicate proximity to a south-polar continental ice cap.

Facies analysis remains one of the geologist's more important occupations; although it is an academic exercise, much of it has been achieved by geologists in the oil and gas industry. Borehole data have greatly extended the exploration geologists' ability to hypothesise and to test ideas on the three-dimensional properties of facies. Petroleum-exploration geologists have also been in the fore-front of developing ideas about sedimentation in relation to tectonics, the subject of the next chapter. For present purposes the grouping of facies and environments is rather broad, as subsequent chapter headings suggest.

3

Stratigraphy and the World Tectonic Model

The character and relationships of sedimentary formations are clues to the roles of sedimentation and tectonism in the geological cycle. The sites of sediment (and volcanic) accumulation—depositional basins—that are preserved in the record are all to a greater or lesser extent determined and controlled by earth movements. In the tectonic framework of sedimentation we see the combination of variously moving or relatively immobile tectonic elements in and around a depositional basin. There are about 600 depositional basins in the world today. They vary greatly in size and shape, in their contents and in the nature of their growth or shrinkage. Within the basin itself the nature and thickness of the sediments may reflect the contemporary tectonic behaviour of the earth's crust locally. The contrast between sedimentary formations of stable shelves and platforms and of active tectonic regions and mountain belts is striking. Isostatic adjustment by uplift following orogeny explains the long-continuing role of orogenic belts as sources of sediment for new basins of deposition, but the initiation of the orogenic cycle has presented a problem. In recent years the new plate-tectonic models of the earth's crustal behaviour have enabled us to understand a possible means by which some of the largest and longest-enduring basins of deposition have originated and evolved.

In the mid-nineteenth century James Hall's study of the Palaeozoic succession of the Appalachian region of North America led him to believe that the weight of the accumulating sediments had caused enough subsidence to allow shallow-water deposition to persist until a thickness of several kilo-

metres had been achieved. J. A. Dana, on the other hand, argued that it was the slow subsidence of a portion of the crust that allowed sedimentation to continue there. He followed this with his classic paper on the origin of mountains and the nature of the earth's interior and introduced the term *geosynclinal* to define 'a long-continued subsidence'. Schuchert (1923), Kay (1951) and many others have reviewed progress in the study of geosynclines and in the period up to the late 1960s a major theme of geological debate and research (see figure 3.1) was the phenomenon of geosynclinal accumulations and their deformation in orogenic belts (Aubouin, 1965).

Kay (1951) estimated that 'geosynclines' have been the major sediment traps on the earth's surface and that 82 per cent by volume of all sedimentary rocks in North America are geosynclinal, with only 18 per cent being cratonic. It has been argued that the ultimate fate of all sediments is to be deposited in the oceanic basins. The length of time they take to reach this environment may vary greatly, and the bulk of terrigenous deep-water sediments are silts or muds. The plate-tectonic process eventually returns some of this material to the craton, as in an orogenic episode at a subduction zone or where continents collide.

With the discovery of sea-floor spreading and of the division of the earth's crust into renewable plates upon which the continents are carried, this debate has become one of adapting the hypothesis of plate tectonics to account for the tectonic settings and the tectonism that influence major depositional areas. Data accumulating from so many different sources now seem to confirm the view held by Dana

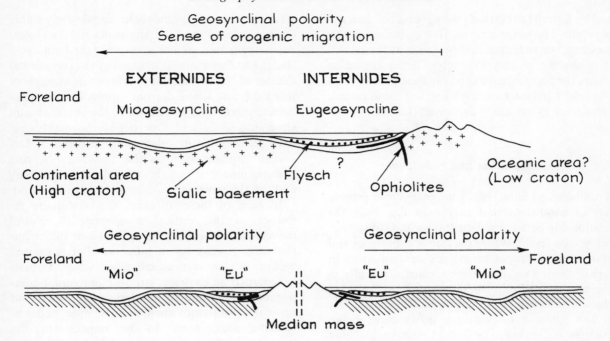

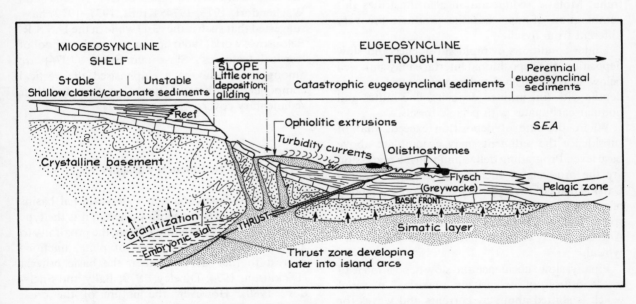

Figure 3.1. (Top) The classical geosyncline was more apparent than real but all so-called geosynclinal belts seem to have possessed some of the essential features shown above. An *orthogeosyncline* consisted of *miogeosyncline* and *eugeosyncline* components, deeper-water facies and volcanic associations being characteristic of the latter. In the upper figures (after Aubouin, 1965) the nature and distribution of different types of crust (continental or oceanic) was not included as being of vital significance but horizontal movement is given great prominence. (Bottom) The concept of plate tectonics was to supplant these models in the mid-1960s; Kuendig's (1959) model draws close to the plate-tectonic concept of a destructive plate margin and includes many features found in the Devonian 'geosynclines'.

and by Barrell (1917) that subsidence is the underlying control on sedimentation. The questions of why it occurs where it does and proceeds at the rates it does are not so easy to resolve. Basin formation within the oceanic realm is relatively simple in terms of present plate-tectonic theory, but intracratonic subsidence is less easily explained (see below).

Sedimentation and Subsidence

Krumbein and Sloss (1963) designated four general sets of conditions that may exist regarding the relationship between sedimentation and subsidence below base level*. Where the rate of subsidence and of deposition is rapid, large thick accumulations of rapidly buried sediments are formed. Typically in the marine realm turbidite sedimentation takes place. The Lower Devonian of the Rhineland, part of the Rhenish or Variscan geosyncline, is an example; others are to be found in the Appalachian region of North America, China and eastern Australia. Molasse sedimentation also illustrates this combination of rates but the products are very different from flysch.

Under conditions of rapid subsidence and slow deposition, the depositional interface may sink well below the regional level and may even reach oceanic depths with fine-grained deposits, fine lutites and nodular carbonates with pelagic fossils.

Where the rate of deposition exceeds that of subsidence the sediments may accumulate above base level. Prograding deltas and marine regression are the products of such activity. The Devonian Catskill delta of the Appalachian region was a body of sediment that rapidly reached base level and then spread laterally westwards from the source area. Arkosic, poorly sorted sands and conglomerates are typical.

Finally, slow subsidence and slow deposition lead to deposition of extensively reworked material which is shifted about by currents and waves for some time before it comes to rest. These conditions

---

* Base level is a concept introduced by Barrell (1917) as the initial level of erosion and deposition represented by river-flood level of the continents and wave or current base in the sea (it is a surface of equilibrium between sediment accumulation and erosional activity).

seem to run most commonly in the relatively stable (cratonic) interiors of the continents and the typical deposit may be a mature well-sorted sandstone or a shale. The Devonian formations of the continental interior of North America provide a fine example of this kind, and where detrital supply was low, carbonate sheets were laid down. In the oceanic realm slow subsidence and slow deposition may occur over sea mounts or other positive features. Devonian deposits such as the ammonoid-bearing nodular limestones on the 'schwellen' of Germany and the Hercynian area of Europe are of this kind.

In each of these instances the continuing influence of the vertical component of crustal movement—subsidence or uplift—is of major importance. It may be an influence that is exerted because of plate-tectonic movement which is usually regarded as having a stronger horizontal component. But several writers have emphasised that vertical crustal movements on a wide and impressive scale seem to be implicit in the Proterozoic–Palaeozoic geology of central Europe. In particular, Krebs and others (Krebs and Wachendorf, 1973, 1974; Krebs, 1977, 1979) have suggested that such is the case, while in the U.S.S.R. Belousov's work (1966) has engendered a school of 'vertical tectonics'. Sloss and Speed (1974) are among those who have emphasised the vertical components in the evolution of cratons and their sedimentary cover.

Sedimentary Basins

In a general discussion major depositional basins may be ascribed to categories wherein it is the type of crust that serves as a foundation; the proximity to a plate margin and the type of plate junctions nearest to the basin are then the basic criteria (Dickinson, 1974; Dineley, 1979; Bally and Snelsdon, 1980). However, the nature of the crust beneath many basins is so far unknown and crustal behaviour may be of greater value as a criterion. Klemme (1980) included the nature of the sedimentary pile, its thermal history and other characteristics of interest to the petroleum geologist in a synopsis of sedimentary basins but the basic criteria for classification are much the same.

Cratonic basins

These are situated within the framework of a craton and developed on continental crust (ensialic), or situated between two recently juxtaposed cratons and developed on continental and transitional crust. Cratonic basins may persist for long periods of time. A relatively small change in base level may profoundly alter the extent of the bodies of water or the sediment that occupies them.

Basin formation well within the confines of the plates and within cratonic areas poses a problem in that it cannot obviously be related to processes that operate at the plate margins. McKenzie (1978) suggested a model in which tension leads to thinning of the crust, followed by tensional faulting and subsequent sinking of the cratonic surface. Bott (1982) has offered a choice of possible mechanisms to account for the origins of these basins: (1) loading that causes flexuring and subsidence; (2) heating of the lithosphere which then thins and subsides as it cools; and (3) subsidence from stress, followed by faulting as the lower crust moves by ductile flow. Probably all three kinds of origin have occurred from time to time. The infilling of the basins by masses of sediment provides a record of the speed and areal extent of the subsidence and detailed analysis of these basins is leading to the view that three important phases in such basin evolution are common. First is a phase of locally confined flexural subsidence. Faulting and an increased pace of subsidence and spreading follows, with a final phase of slower subsidence and great lateral spreading of deposition.

Extracratonic basins

These are situated on the flanks or at the margin of a craton and underlain by continental, transitional or (rarely) oceanic crust. Here again, the nature of what underlies the basins at the *margin* of a craton is commonly unknown. This group is subdivided into: (a) passive marginal basins, Atlantic or discordant continental margins (Dewey and Bird, 1970; Hsü, 1972; Bott, 1980); and (b) active marginal basins (subducting margins), Andean and Western Pacific or concordant continental margins (Dietz and Holden, 1966; Mitchell and Reading, 1969).

Oceanic basins

These are situated on oceanic (ensimatic) crust between ridge and arc, trench or continental margins.

In each of these basins distinctive rock assemblages are present and the literature on the relationship of sedimentary and volcanic suites to the (plate)-tectonic situations in which they occur is now large. Dickinson (1974) and others have given many examples that are related to the simple models shown in figures 3.2 and 3.3.

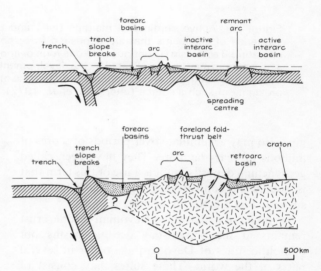

Figure 3.2. The basic elements of basin formation at active continental margins as visualised by Dickinson (1974). Deposition is vigorous and may be supplemented by volcanic accumulations and igneous intrusion. Crustal uplift, fracture and warping may return or change location from time to time.

Basins that are associated with arc–trench systems may either be comparatively short-lived or may exist for immense lengths of time. Similarly, many individual basins that are associated with continental collision may be relatively short in duration while others persist longer. There is a natural tendency for the basins here to change vigorously in shape, size and continuity as the collision proceeds. Prisms of sediment on the continental shelves, slopes and rises along passive mar-

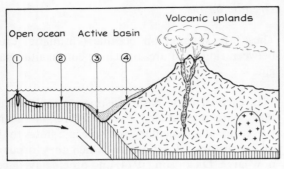

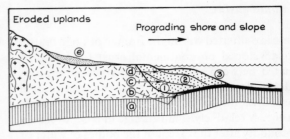

1. Basalts extruded

2. Deep water sediments (pelagic)

3. Pelagic sediments overlain by detritus from continent

4. Greywackes

a) Submarine lavas

b) Carbonates or siliceous oozes

c) Thick clastics, turbidites, etc.

d) Neritic carbonates and clastics

e) Molasse and terrestrial deposits

Figure 3.3. Active continental margins (left) and passive margins (right) each have a characteristic sequence of sediments that are deposited in response to the tectonic setting and surface processes operating. Tectonism is continuous in one form or another on the active margin but rare on the passive. The active margin may be complicated by the development of island arcs; the passive margin may be adjacent to or far from an active source of sediment and the rate of growth of the offshore clastic wedge may reflect the proximity of such a source (after Gass *et al.*, 1972).

gins also may accumulate over periods of scores of millions of years or more. They have been called geosynclinal deposits but it is a use of this term that is somewhat controversial; 'geoclinal' has been suggested as a better descriptive term.

No intact or undeformed Devonian oceanic crust remains in the present-day oceanic basins but ophiolitic suites of Devonian age occur in several parts of the world. These suites may consist of serpentinised peridotite and dunite, gabbro and sheeted dyke complexes and pillow lavas topped by sediments. Pelagic fine-grained rocks, black shales and cherts, or less commonly neritic sedimentary rocks, are often found in association with the igneous materials but intense deformation may obscure the relationships. Only in areas of severe orogeny or unusually severe displacement of parts of the crust can ophiolitic materials be raised from their deep site of origin, but the processes involved commonly render an interpretation of the original materials very difficult.

While plate-tectonic theory provides models in which all manner of sedimentary environments may be possible, it is commonly quite difficult to relate a particular episode or setting of sedimentation to plate-tectonic processes. So far, for example, no

completely convincing plate-tectonic explanation of the Rhenohercynian basin and its evolution has been produced. Sedimentary facies are in part only at the end of a chain of processes that are involved with plate evolution. After all is said and done, it is what transpires in the lithosphere that creates the tectonic setting for sedimentation and what occurs within that setting is influenced by hydrosphere, atmosphere and biosphere to a very large degree. We can observe and understand these *external* agencies to a very much greater extent than we can observe processes within the lithosphere, the *internal* agencies.

Tectonic settings immediately impress us by the variable rates and magnitude of crustal mobility. Mobile zones especially have seemed to be of three kinds: compressional, with folds and thrusts; extensional, with normal faults and hypabyssal intrusions and volcanoes; and horizontal shear zones with transcurrent strike–slip faults. The place for each of these within plate-tectonic theory is clear and unequivocal—convergent plate margins, divergent plate margins, and transform margins where one plate grinds past another without any substantial addition or removal of the lithosphere. Plate margins are thus seemingly all important.

For the most part this is acceptable enough, if difficult to apply to many areas, but sedimentary basins do develop at great distances from plate margins. Remarkable basin developments may also occur within cratonic interiors. These may be regarded as the side-effects of orogeny at continental margins in some instances, but the mechanisms that lead to cratonic basin formation have commonly still to be demonstrated in more than a most general way (Bott, 1976; McKenzie, 1978).

Mitchell and Reading's (1978) rather similar synopsis of plate-tectonic settings for sedimentary basins postulates the following settings (see also Bally and Snelsdon, 1980):

I. *Spreading-related.* (a) Intracontinental rifts, (b) failed rifts — long-lived deep linear troughs or *aulacogens*, (c) intercontinental rifts that give passive continental (Atlantic) type margins and basins.

II. *Subduction-related.* (a) Fore arc basins, (b) ocean floor, deep-sea trenches and outer arcs, (c) outer arc troughs, (d) volcanic arcs, (e) back arc areas.

III. *Transform strike–slip fault-related* (single-fault linear basins).

IV. *Continental collision-related* (see figure 3.4). (a) Remnant ocean basins, (b) late orogenic basins.

Many ancient analogues for these can be found: in the geology of the continents they underline the fact that continents themselves are relatively passive objects except to a small extent where the forces of the ocean crust are released against their margins.

Subduction and strike–slip movements produce

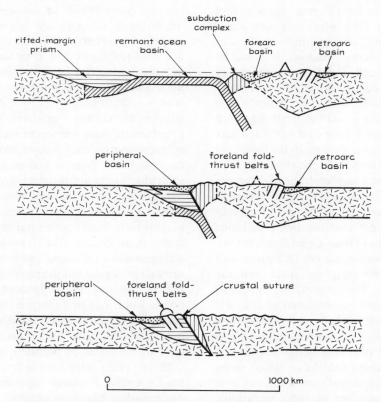

Figure 3.4. Perhaps the maximum tectonic effects are achieved on the collision of continents (Dickinson, 1974). Here flysch sedimentation is succeeded by molasse. Hypabyssal extension is joined by plutonic intrusive activity and isostatic uplift may occur vigorously and over wide areas.

localised surface expressions of the crust that in turn generate very particular and special rock facies. The stratigrapher or sedimentologist may not recognise these facies for what they are. Friend (1981) noted extensive high-angle contemporaneous faults in many Devonian basins in the North Atlantic borderlands, trending parallel to folds in the pre-Devonian basement and indicating horizontal (strike–slip) motion. He suggested that basin formation within the Caledonide orogenic belt was associated with deep faults, perhaps associated with the main collision zone: external basins appear not to reflect such a fault-influenced origin. However difficult it is to interpret Devonian mobile belts in terms of modern plate-tectonic movements, it is thought unlikely that these phenomena have changed very much since then. Subsequent events may, nevertheless, have greatly marred the record. Sedimentary basins and their contents do depend upon the interactions of both internal agencies and external processes in their evolution and many geologists are concerned with the history of basin infillings for purely economic reasons. It would be rash to suggest that all Devonian sedimentary basins can be relegated to one or other of these divisions but many can be seen as examples of one or other of them. So far it has not been suggested that the Devonian was a time that favoured any particular kind of basin formation. In the future we may, however, be led to a different view.

Apparent in all these considerations of sedimentary basins is the need to categorise the wide variety that seems to exist (see figure 3.5) and to explain their existence under different tectonic conditions and for different lengths of time. Crustal movement, both vertical *and* horizontal, has been the prime and ultimate agency in providing the space and the infilling of these basins. The more extensive, thicker bodies of sediment have accumulated at cratonic margins and over wide seafloors or oceanfloors. Many of them are now preserved as more or less deformed prisms at or near continental margins or squeezed between cratons that have closed upon them. Masses of Devonian sediments and volcanic rocks occur in one or other of these situations, having been involved in sedimentary basins that existed for long periods of geological time. They also occur in very large continental basins that formed subsequent to the deformation of the mar-

ginal long-lived basins, 'the geosynclines'. And in yet a third category they are conspicuous members of the very shallow, very broad 'basins' that existed when epeiric flooding of the craton surfaces occurred. Here again tectonic activity is not to be ignored. It was certainly the cause of the formation of short-lived gentle basins and troughs, domes and arches that affected these otherwise singularly unrelieved surfaces.

Cratonic Sedimentary Sequences

During the mid part of the present century the intensive geological exploration of the continents, especially subsurface exploration, has revealed details of the extensive transgressions and regressions that have affected the continental interiors. This has been especially the case in North America where it enabled Sloss (1963) to distinguish a number of 'major rock stratigraphic units of inter-regional scope', separated one from another by unconformities (see figure 3.6). Over the major part of the continental interior he distinguished six such units in the succession from Late Precambrian to Recent, with a seventh regression now in progress. At the edges of the craton, Appalachian, Cordilleran and Franklinian, these sequences merge with the thicker, more continuous, geosynclinal kind, but within each sequence there is a strong homogeneity and individual units are traceable over wide areas (see figure 3.7). The bounding unconformities are similarly traceable over great distances and show rising diachronism from continental margins towards the continental centre. The transgressive phases of sedimentation following upon each unconformity are well preserved but the regressive phases are only poorly retained in the record. Johnson (1971) pointed out that non-marine rocks should not be included in these sequences and emphasised that in many areas the latter are not so much separated by unconformities as by marine hiatuses.

Sloss (1972) later turned his attention to data from the great Eurasian continental land mass and distinguished there a pattern of unconformity-bounded cratonic stratigraphic units of a similar kind. It is difficult to avoid the idea that these cratonic sequences are represented on most of the continents as a result of world-wide rise and fall of

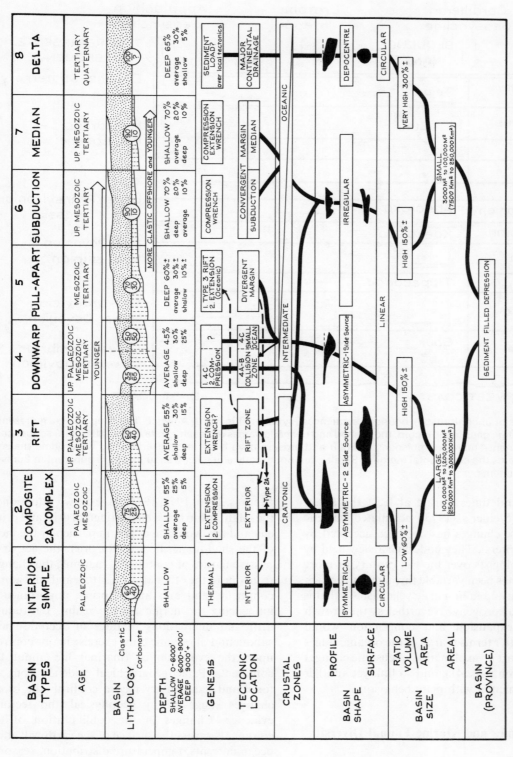

Figure 3.5. The enormous variety of sedimentary basins is the subject of much study by petroleum geologists. Here is one classification which utilises many features other than the essentially tectonic. The tectonic location refers to the basin's position relative to the craton margin or plate margin (after Klemme, 1980).

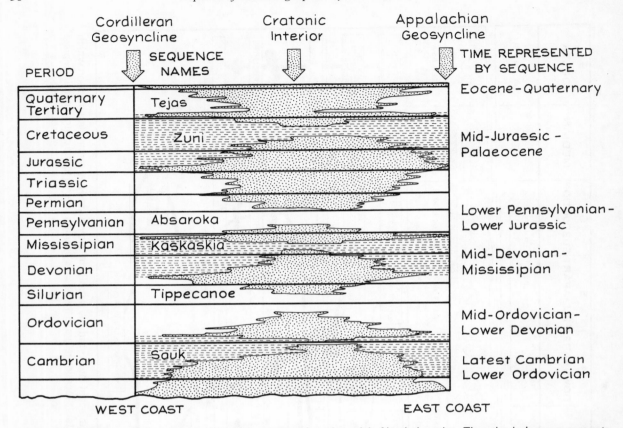

Figure 3.6. Major unconformity—bounded sequences (synthems) in North America. The stippled area represents significant or large gaps in the record; the dashed and unshaded areas represent strata of the (alternate) sequences. Thus six continent-embracing episodes of uplift and erosion have been recognised during which time the sea withdrew towards the continental margins. Similar, but not exactly equivalent, sequences are apparent in Eurasia and other continents.

sea level several times during at least the Phanerozoic eon (Steiner's (1973) postulated galactic cause for these changes has not been accepted by subsequent writers). They underline a view held by numerous geologists over the last century or more that sea level has been variable throughout geological time, rising and falling in a cyclic fashion that is independent of orogeny. Nevertheless the growth and decay (expansion and contraction would perhaps be better terms to use) of the oceanic ridges may offer in part an explanation of eustatic changes (as described on page 41), and so link yet another global phenomenon with plate behaviour.

Plate Tectonics and Marine Faunal Diversity

Several authors (for example, Valentine and

Moores, 1970, 1972, 1974) have drawn attention to the relationship between continental drift and orogeny and the evolution of life, in particular the biota of the shallow neritic seas. In brief, the active or passive nature of continental margins, the changes from one to the other and the rifting and collision of continents and arcs are of prime importance in influencing the neritic marine realm. Latitudinal drift and fluctuations in rates of accretion and subduction of oceanic crust exercise major controls upon the evolution of (marine) life about the continents. Palaeogeographical evidence from the fossil and sedimentary record confirms the broad changes suggested by geophysical and tectonic evidence. Changes in the configuration of the continents across the surface of the earth will affect ocean currents, temperature distribution, seasonal fluctuations and the distribution of nutrients, pat-

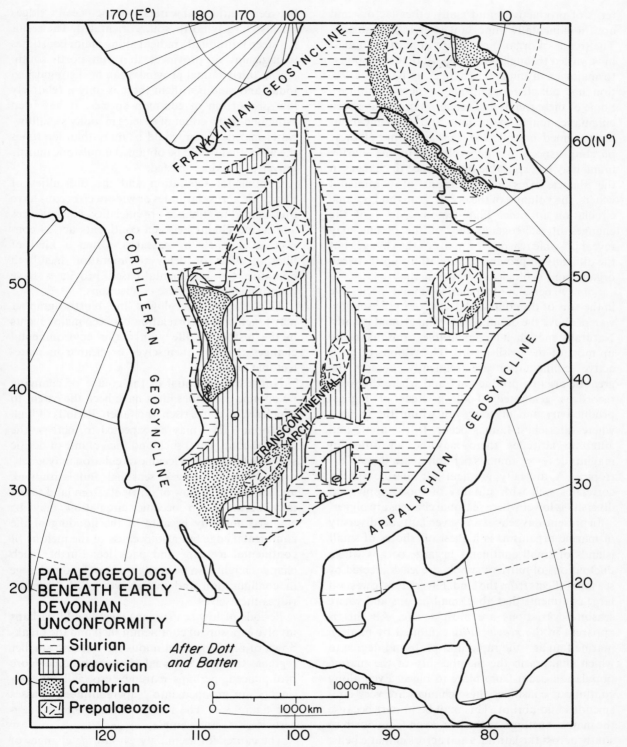

Figure 3.7. The Kaskaskia sequence rests upon a warped surface that involves Precambrian to Silurian rocks. At its outset the geology of the North American craton may have been as shown above (after Dott and Batten, 1971).

terns of productivity and many other factors that are fundamentally important to living organisms. The trends of organic evolution in marine animals have varied throughout geological time in response to major environmental changes as natural selection has adapted organisms to new conditions. Some of these trends and faunal changes have been enigmatic; major extinctions and diversifications have seemed inexplicable. Perhaps some of these puzzling aspects of the evolution of the marine fauna may be explicable by reference to changes in the shapes, sizes and positions of continental waters, the volumes of the ocean basins and oceanic circulation in time. Plate-tectonic views of the changes in palaeogeography of the oceans may reveal possible reasons for many of the changes but the difficulty of proving such relationships between organic and geological events remains.

Some 90 per cent of all marine animal species live in the seas of the continental shelf or in the shallow waters above the oceanic 'rises' and around islands. Naturally enough, it is comparable forms that make up most of the fossil record. The regions of most active sedimentation are frequently those of the greatest organic productivity and where the fossil record is generated. Today's areas of highest productivity and diversity are within the tropics, where the richest faunas occur. In the arctic and antarctic latitudes there may be an order of magnitude fewer animals per unit of sea than in the tropics. A diversity gradient for marine species correlates well with stability of food supply and diversity is lower where seasonal changes are bigger.

In present-day seas at any given latitude, diversity of marine organisms is highest off shores of small islands or small continents in large oceans where fluctuations of nutrient supply are least affected by seasonal effects from the land. Diversity is lowest off large continents that abut small oceans and where seasonal variations are most severe. All this is apparent in the *provinciality* exhibited by modern marine faunas — the rapid and profound degree to which changes in the membership of the marine population occur from place to place. Even along continuous coastlines these changes are very conspicuous and abrupt. This may be in part because the major coastlines of the world run from north to south, across the latitudes and across climatic belts. The ocean deeps constitute an effective barrier between continental waters, but the oceanic ridges are also largely north–south orientated. The result is a steep latitudinal gradient along major coastlines with chains of provinces that run north–south. About thirty faunal provinces can be distinguished today and amongst them there is only a relatively low proportion of common species. It has been estimated that a count of species in today's shallow-water marine fauna would be more than ten times the value that would be obtained if only one marine universal province existed.

Continental separation and the difficulties of migration across latitudes or wide deep oceans have thus tended to enhance provinciality; volcanic arcs and island chains between continents act as connecting channels or passages. Often a kind of filtering effect or delay to migration has been operative but continental-based provinces have been influenced by these connections.

Thus the palaeoecologist and biostratigrapher may together find correlation between major events in the history of life with major environmental changes that are themselves explicable in plate-tectonic terms.

Not least influential in the course of changing taxonomic diversities must have been the extent to which sea level has risen and fallen. Periods of high sea (ocean) level may be expected to increase the areas of continental shelf and the volume of neritic seas (see figure 3.8). Oceanic circulation may ameliorate climatic changes, seasonal and latitudinal, and an improved flow of nutrients from land to sea may occur. Deep oceanic circulation may be changed not only because of the flooding of the continental edge but also because of the melting of continental ice caps and pack ice. Further such changes might also be brought about by the change in configuration and the growth of the oceanic ridges themselves.

Periods of low sea (ocean) level would perhaps involve exposure of continental shelf and the breakdown of one large continuous habitat into smaller regions. Climatic seasonal changes become more pronounced, perhaps extreme, especially if continents are grouped into 'supercontinent' masses (see figure 3.9). The Permo–Triassic interval suggests itself as an example.

The causes for seemingly *world-wide* changes of ocean level, eustatic changes that is, have been

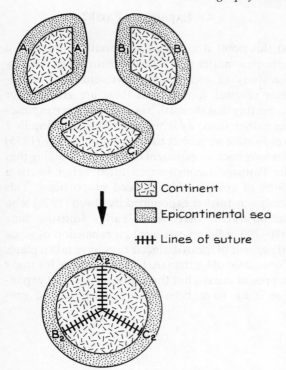

Continent

Epicontinental sea

┼┼┼ Lines of suture

$A_1 + B_1 + C_1 > A_2 + B_2 + C_2$ Lower provinciality
Higher continentality ∴ Lower diversity

Figure 3.8. Changes in diversity of the biotas of the marine neritic environments brought about by the gathering of several smaller continents into a supercontinent as postulated by Hallam (1974).

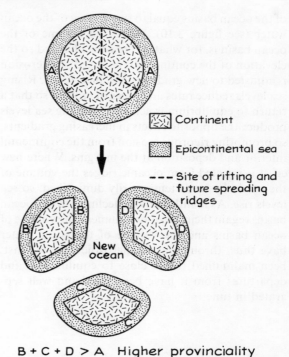

Continent

Epicontinental sea

--- Site of rifting and future spreading ridges

$B + C + D > A$ Higher provinciality
Lower continentality ∴ Higher diversity

Figure 3.9. When a supercontinent splits into several smaller continents there is an increase in the size of the epicontinental seas, thus allowing an increase in diversity of environments and organisms (Hallam's model of 1974).

debated (Hallam, 1974; Hays and Pitman, 1973; Donovan, D. T. and Jones, 1979) and a growing consensus favours the changes in ocean-ridge volume as a control. Times of increasing production of oceanic crust cause an expansion of the oceanic ridges by accretion, intrusion and rise in temperature. Times of quietude or inactivity allow the ridges to subside by tectonic movement and thermal contraction. Vigorous ridge activity and growth may be so prolonged as ultimately to bring about relatively sudden drastic events such as continental collision.

The pattern of subduction and the growth of new ocean crust may thereupon change, with attendant effects upon eustatic sea level. It is doubtful that the evolution of pre-Mesozoic oceanic ridges will ever be understood well enough to establish convinc-

ingly their essential role in regulating the worldwide changes of sea level of those times. Local isostatic changes as a result of orogeny and 'unloading' of cratons also tend to add complicating factors to the model. Indeed, whatever the intermittent activities of the oceanic ridges, the continents themselves seem to have maintained a fairly constant 'freeboard', or elevation relative to sea level, throughout the Phanerozoic. Wise (1974) believes that for some 80 per cent of the Phanerozoic eon the freeboard has remained within 60 m or so of a horizon that is about 20 m above present sea level. He sees the continental margins as the regions wherein sediments that derive from the continent are welded back on to the sialic crust. His global tectonic model provides constant continental freeboard, and mechanisms that maintain the volume of the ocean basins equal to the volume of the ocean water (see figure 3.10). When the volume of the ocean basin is, for whatever reason, changed so the elevation of the continents is changed, and erosion is adjusted to new gradients and base levels. Rising sea levels reduce rates and areas of erosion so that a return to equilibrium is achieved. Falling sea levels produce the opposite effects in increasing gradients, so promoting rigorous erosion from the continental interior and deposition at the margins. Where new crust is injected from oceanic ridges the volume of the ocean basins is temporarily diminished, so sea levels rise. As mantle activity declines, so the ocean basins regain their previous volumes. The volume of ocean basins and the volume of the ocean water have thus, throughout Phanerozoic time at least, been maintained rather close to equilibrium, and departures from it have been brief and well separated in time.

An Expanding Earth?

At this point it may be appropriate to interject a note on a matter that is thrown into sharp relief by the study of continental movements and ocean-floor renewal, namely that there are grounds for suspecting that the earth has with time been expanding rather than, as is usually thought, being in a steady state or subject to contraction. Carey (1975) has long been an exponent of this view, holding that the Pangaea reconstruction fitted better on to a globe of considerably reduced proportions. This thesis was further expounded by Owen (1976) who demonstrated that since early Jurassic time (180–200 million years ago) an expansion of some 20 per cent of mean diameter may have taken place. Ocean floor older than that is not available for study in present oceans but the implication is that expansion may have been taking place in the pre-

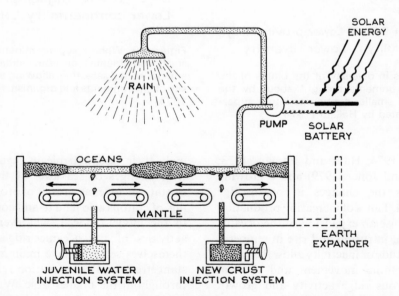

Figure 3.10. The model proposed by Wise (1974) for a possible constant freeboard global tectonic model with feedback mechanisms that maintain the volume of ocean basins equal to the volume of ocean water. The control of juvenile water and of new crust injection is in fact shared for the most part. The rate at which the rain is discharged is controlled by solar radiation (assumed to be constant) and by the size (area) of the oceans. Increased area results in increased precipitation and hence increased erosion. Equilibrium is maintained by the interaction of mantle activity and surficial processes.

Mesozoic. Several ideas about the reason for an expansion of this magnitude have been put forward but Owen finds most favour for relating this trend to that of the general expansion of the universe now admitted by astronomers. In 1981 Carey's ideas, now embodied in a large book, were the instigation of an international gathering in Sydney (see also Ollier, 1981). There the very existence of subduction zones was called into question—the zones exist, but not for subduction according to Carey, quite the opposite. The opponents of his earlier ideas have not been able to demonstrate in practice that palaeomagnetic evidence refutes the hypothesis. New means are currently being devised to test it but we must not expect immediately convincing results one way or the other. Explanation of an *expanding* earth is even more difficult than demonstrating its possible existence but phase change at constant mass, secular decrease in *G* and secular increase in mass have all been invoked by one writer or another. The implications of an expansion of this kind for gravity and for the atmosphere, oceans and the biosphere could be significant but must remain conjectural for now.

Oceanic and Atmospheric Repercussions

Fischer (1979) has found in mantle convection a motor mechanism not only for plate tectonics and the ultimate cause of volcanism and continental freeboard but also for atmospheric carbon dioxide content. The relatively youthful discipline of sedimentary geochemistry and the extensive recovery of oceanic cores has provided data that suggest significant geochemical cycles may have operated at length and were dependent upon crustal events. Fischer sees rhythmic changes in the structure and behaviour of the atmosphere and the oceans resulting from changes in the intensity of mantle convection throughout the last 700 million years. A 300 million year cycle, he views as having occurred twice in this time, alternating between two states and modulated by shorter cycles.

Briefly, the two states are:

O State. Low mean sea level, continents high and mountainous and with episodic cover by ice sheets. Latitudinal temperature gradients steep, vertical gradients in seas strong and oceans vigorously convective, their waters strongly oxygenated. The O State has a low proportion of atmospheric carbon dioxide and is one where radiation loss occurs.

G State. High mean sea level, continents low, latitudinal temperature gradients gentle, oceanic vertical temperature gradients also gentle with the oceans sluggish and prone to anoxia. Here the atmospheric carbon dioxide content is high and produces a 'greenhouse effect', either warm or cold. Continental glaciers form only in the extreme G State.

Transitions between these states involve major changes to existing climatic patterns and resulting crises for life. These are seen as occurring at 150 million year intervals.

Siluro–Devonian times were given over to the G State and a transitional phase to O State was established in the Late Devonian. The consequences of this for the stratigraphic column in terms of the sedimentary formations and biota that were developed during the Devonian period are discussed in chapter 10.

Debate about plate-tectonic processes and their effects is still vigorous but it is generally accepted that the principle of plate movement is established. The simple models that were set up to explain major ancient and modern crustal features will need elaboration as new data are acquired (see figure 3.11). It is attractive to believe that atmospheric, climatic and oceanic evolution has been linked via volcanic activity to the same mechanism and that the major world environmental changes are reflected in organic evolution. There are many uncertainties as yet unresolved. Several interpretations, expressing very different views of the structure and evolution of the best-known orogenic belts, have to be reconciled. Many geologists are as yet unconvinced that we shall ever gather sufficient hard evidence to prove hypotheses about atmospheric or oceanic evolution. Nevertheless, plate-tectonic theory offers a foundation upon which explanations of long-continuing geological and biological activity may be set. Increasing knowledge of the evolution of the crust since Pangaea began to break up allows us to postulate more convincingly the processes that gathered in the dispersed Palaeozoic continents and regulated the Palaeozoic environments.

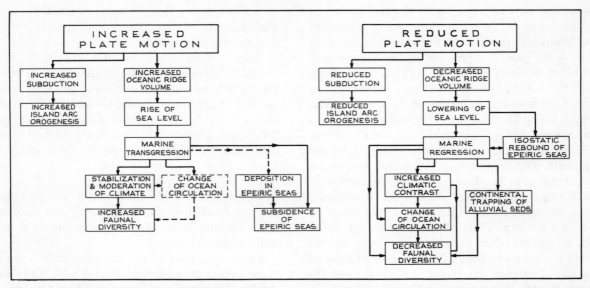

Figure 3.11. Hays and Pitman's (1973) proposed sequence of events caused by an increase in the speed of plate movement and a subsequent decrease. Originally postulated for the Upper Cretaceous transgression it also holds good for other transgressions such as that of the Late Devonian.

Sedimentary basins that contain Devonian rocks (though not necessarily exclusively) can be found for almost every basin category listed in this chapter. As stratigraphic analysis proceeds the local crustal mechanisms may be indicated sufficiently well to allow comparisons with later and better-understood instances. The relatively large volumes of sedimentary and volcanic formations that were generated in the Devonian period indicate both crustal and surficial activity at a varying rate but are generally greater than for the other Palaeozoic periods. The variety of basin settings and structures may confirm this.

In this chapter we have considered the possible tectonic mechanisms that influence the accumulation and preservation of stratigraphic successions. The plate-tectonic model has now been used in very many different instances to explain the origin and development of sedimentary basins and their infillings. While many of these are convincing, others are not, and new locally more complex mechanisms must be sought in explanation. Despite the bur-

geoning literature on the subject, plate-tectonic theory remains a relatively crude means of solving problems of crustal history, and the refining of ideas has to continue. Moreover, the plate-tectonic model has some far-reaching implications for palaeontology in terms of geographical change. These too need careful evaluation and testing. Phenomena such as magnetic reversal, change in the length of the day and year, terrestrial expansion or contraction, the value of G and other features may be related in that changes deep within the mantle and core are their cause. One is reminded of the first 'law' of ecology, namely that 'everything is connected to everything else'. The consideration of how these possibly related phenomena—if they are indeed real and not imaginary—and models affect our view of the earth in Devonian time is full of uncertainties, but the exercise should lead us to accept that it was probably in more ways different from the present than we had thought. The case will be deliberated further in chapter 10.

4

Time in Question: the Devonian

The second quarter of the nineteenth century was a time of phenomenal achievement in stratigraphical description and attempts to establish the major train of events in the history of the earth. The highly fossiliferous Mesozoic formations of Europe and southern Britain were already well known and the Carboniferous system, so important to the industrial revolution then in progress, was also recognised as a major stratigraphical unit. The Old Red Sandstone of Scotland, Ireland and South Wales was regarded as the basal formation of the Carboniferous. The order of formations beneath the Old Red Sandstone, however, was much less certain. Crystalline rocks were encountered in some areas, and elsewhere limestones and other fossiliferous beds, little deformed, could be recognised. The picture was confusing and in the early 1830s the rocks beneath the Old Red Sandstone were generally known as 'Greywacke' or 'Transition' formations.

T. H. De La Beche had been commissioned by the British government of the day to prepare a geological report on the south-west of England and by 1836 his interpretation of the geology of Devon and Cornwall had become well known enough for some fairly sharp debate about it to be taking place at meetings of the Geological Society of London and at the annual meetings of the British Association. The principal argument concerned the structure of the region and De La Beche was under fire from the energetic Roderick I. Murchison and Adam Sedgwick (Rudwick, 1979). He came to accept their different interpretation of the structure of Devon and agreed that the Culm Measures were the youngest beds beneath the New Red Sandstone, but he held that the Culm Measures were conformable upon rocks that Murchison claimed as Lower Silurian. He suggested that perhaps the Culm would prove to be Upper Silurian in age rather than Carboniferous.

It was at that time believed that the faunas of the Carboniferous Limestone were the immediate and direct successors to the Silurian faunas that were discovered by Murchison in Wales and the Welsh borderlands. During the period 1837–1840 a number of amateur geologists had become familiar not only with the fossils of the Carboniferous Limestone and the Silurian of the Welsh borderlands but also with those from the limestones of south Devon. They commented on the similarities and differences and in 1839 Sedgwick and Murchison published the view that the rocks and fossils from Devon indicated a new system that lay between the Silurian and the Carboniferous. No mention was made of the fact that the Old Red Sandstone was the *non-marine* equivalent of the Devonian strata south of the Bristol Channel.

In 1839 Murchison visited the splendid sections through the Devonian of the Rhineland and in that same year with help from the palaeontologists Phillips, Sowerby and Lonsdale he felt he had firmly established the palaeontological and stratigraphical identity of the Devonian system. Shortly afterwards other geologists from Europe were to confirm his findings there and to establish that the Old Red Sandstone is indeed the equivalent of the Devonian. It was, interestingly, at about this time that the Swiss geologist Gressly introduced the concept of facies—'the total complex of all primary lithological and palaeontological characteristics of a sedimentary unit' (see page 27).

So thorough were the accomplishments in this early phase of Devonian stratigraphic work that at the end of it a terminology of series, stages, *Stufen* (see page 48), and other subdivisions had become accepted for the system as a whole. During that 30 years the geology of the Ardennes–Rhineland area of the Devonian was mapped in outline and the early claims that it contained a full succession that was equivalent to the Silurian, Devonian and Carboniferous of Britain were substantiated. The work of E. Kayser, J. Gosselet, C. Koch, H. Dorlodot and F. Frech (see figure 4.1) was of importance to W. A. E. Ussher of the Geological Survey of Great Britain when he was sent into Devon to map the region for the New Series One-inch Geological Map. He was able to bring to Devon the continental scheme of stratigraphic succession for the Devonian rocks (see figure 4.2) and it was immensely helpful to him in ground where the rocks were only locally fossiliferous, or were deformed and interrupted by many dislocations.

During the period that ended just after the second world war an enormous wealth of stratigraphic and biostratigraphic data was collected (see figure 4.3).

	Dumont, 1848	Kayser, 1870, 1871			Gosselet, 1860–1888	Kayser, 1881–1885	Dorlodot, 1900
LOWER DEVONIAN	Système EIFELIEN	VICHTER Schichten (= Schichten von BURNOT)	COBLENZIEN	Supérieur	Graywacke de HIERGES (including beds with *Spirifer cultrijugatus*) Poudingue de BURNOT	COBLENZ-Schichten (Coblenzian)	EMSIEN (Ems-Stufe)
	AHRIEN	Schichten von AHRIEN			Grès de VIREUX		
	supérieur = Hunsrückien COBLENZIEN inférieur = Taunusien	COBLENZ-Schichten (Coblenz-Schiefer)		Inférieur	Graywacke de MONTIGNY Grès d'Anor	SIEGEN-Schichten (Siegenian)	SIEGÉNIEN (Siegen-Stufe)
	supérieur GEDINNE inferieur	Schichten von GEDINNE	GEDINNIEN		Schistes de ST. HUBERT Schistes bigarrés d'OIGNIES Schistes de MONDREPUIS Arkose de HAYBES Poudingue de FÉPIN	GEDINNE-Schichten (Gedinnian)	GEDINNIEN (Gedinne-Stufe)

Figure 4.1. Some of the early researches into Devonian stratigraphy in Western Europe show a strong measure of general agreement on the correlation and subdivision of the Lower Devonian. The fossil content of these rocks is not everywhere abundant nor of great biostratigraphical use, consequently the definition of stages by different workers has not always been consistent or clear (after Ziegler, W., 1979).

		Ardennes		Eifel Hills		East of the Rhine	
	Gosselet, Dewalque	Maillieux, 1927, 1938		Kayser, 1871	Struve, 1961	Kayser, 1890 Frech, 1886	modern
Givetian (Middle Devonian)	Etage du Calcaire de Givet	Gid / Gic	Calcaire à *Hexagonaria quadrigeminum* Schistes et Calcaire argileux à *Spirifer mediotextus*	Upper (*Stringocephalus* Beds)	Schönecken-Dolomit / Bolsdorf / Kerpen	Reef Limestone (*Stringocephalus* Beds)	Massenkalk
		Gib	Calcaire à *Str. burtini* Schistes et Calcaire à	Lower	Rodert / Dreimühlen	Upper / Lenne Slate	Finnentrop Beds
		Gia	*Spirifer undiferus*	Crinoid Bed	Cürten Loogh		Wiedenest Odershausen
Eifelian / **Couvinian** (Middle Devonian)	Etage de Schistes à Cal-céoles (Calcaire de Couvin)	Co2d	*Spyroceras nodulosum*	Upper (Calceola Beds)	Ahbach / Freilingen Junkerberg / Ahrdorf	Lower (Lenne)	Selscheid
		Co2c	*Spinocyrtia ostiolata*				------
		Co2b	Stromatopores (Reefs)		Nohn		Mühlenberg
		Co2a	*Euryspirifer intermedius*	Lower		Lenne	
							Hobräcke

	Zone à *Spirifer cultri-jugatus*	Co1c	*Terebratula loxogonia*	cultri-jugatus Beds	Lauch	Calceola Beds	*cultri-jugatus*
		Co1b	*Uncinulus orbignyanus*				------
		Co1a	*St. piligera* / *A. alatiformis*				Remscheid

Left margin labels: Middle Devonian, Lower, Emsian; Assise de Couvin; Assise de Bure.

Figure 4.2. The Middle Devonian series of Belgium and Germany are highly fossiliferous but laterally very variable. There has in consequence been some disagreement about correlation from one outcrop to another. The stage names refer to Belgian and German sections wherein different precise boundaries have been used by subsequent geologists. As with other series or stage boundaries within the system, these are now being scrutinised by the international (I.C.S.) Subcommission on the Devonian System (after Ziegler, W., 1979).

	Belgian Subdivision Conseil géologique (Maillieux)		Germany Kayser, 1973	Ammonoid-zones based on Wedekind, 1917, and subsequent authors		German Stufen
Famennian	Tn1a	Etroeungt	Clymenia-Stufe	Wocklumeria	*Cyrtoclymenia euryomphala*	Wocklume- rian
	Fa2d	Comblain-au-Pont			*Wocklumeria* VIβ *sphaeroides* *Kalloclymenia* VIα *subarmata*	
	Fa2c	Evieux		Clyme- nia-S	*Gonioclymenia* Vβ *speciosa* *G. hoevelensis* Vα laevigata	Das- bergian
	Fa2b	Montfort		Platycly- menia-S	*Platyclymenia* IV *annulata* *Prolobites* IIIβ *delphinus* *Pseudoclymenia* IIIα *sandbergeri*	Hembergian
	Fa2a	Souverain-Pré				
	Fa1c	Esneux	Cypridina-Stufe	Cheiloceras-Stufe	*Sporadoceras* IIβ *pompeckji*	Nehdenian
	Fa1b	Mariembourg *R. dumonti*			*Cheiloceras* IIα *curvispina*	
	Fa1a	Senzeilles *R. omalusi*				
Matagne	b	Black shale with *B. retrostriata, C. tumida*			post Iδ *Crickites holzapfeli*	
	a	Green shale with *C. armata, B. palmata*			*M. adorfense* Iδ	
Frasnes	i	Shale with *Cyrtospirifer pachyrhynchus*. Bioherms	Intumescens-Stufe	Manticoceras-Stufe	*Manticoceras cordatum*	Adorfian
	g	Limestone with *Cyrtoceras*				
	f	Shale with *Camarophoria megistana*			*M. carinatum* Iγ	
	e	Shale with *Leiorhynchus formosus* and Bioherms				
	c	Shales and limestones with *Pentamerus brevirostris*				
	b	Shales with *C. bisinus, Recept. neptuni*				
	a	Calcareous shale with *Cyrtospirifer orbelianus*			*M. nodulosum* Iβ	
Fromelennes	c	Shale and limestone with *Myoph. transrhenana, L. gilsoni*				
	b	Biohermal reefs with Stromatopores			*Pharciceras lunulicosta* Iα	
	a	Calcareous shale with *Cyrtospirifer tenticulum*				

Figure 4.3. Historical subdivisions of the Belgian and German Upper Devonian sequences (after Ziegler, W., 1979).

Comparisons with other regions of Europe and North America were made. The equivalents of the Devonian series, stages and Stufen were identified in many parts of the world, and the patterns of facies changes from outcrop to outcrop were plotted. Phylogenetic changes in different marine fossil groups such as spiriferids, trilobites and ammonoids were documented and pressed into stratigraphic use. More recently other fossils have been studied in this context, none being more generally useful than the conodonts.

Devonian Marine Facies

The Devonian rocks of the Harz Mountains had during this time been recognised as possessing different characteristics from those of the *Rhenish* facies in the Rhineland and those to the west. Today the facies term *Hercynian* has come to be applied to rocks with oceanic features — reduced rates of sedimentation, lack of terrigenous material and an abundance of fine carbonates and argillites. The fauna is dominantly pelagic with little benthos represented. This facies appeared first in the Lower Devonian of the Harz Mountains and in Czechoslovakia (Bohemia) but spread westwards from Middle Devonian times, reaching Belgium in Late Devonian (Frasnian) time.

In North America the early geological mapping of the central Appalachians and New York State had revealed the extent of the system there. Type localities for the subdivisions of the Devonian for the continent were provided from the highly fossiliferous strata of New York, local names thus indicating the groupings employed (see Cooper *et al.*, 1942; Rickard, 1975). Three series were established, but the Upper Devonian is seemingly thicker than the other two.

The recognition of these stratigraphic units in eastern North America came soon after a European standard subdivision seemed to be established. However it was soon apparent there were different practices even amongst European geologists as to where *exactly* they drew their stratigraphical boundaries and how they correlated between the formations in the different facies. Since the 1950s these difficulties have become better understood and overcome. American geologists have recently

tended to refer their successions more and more to European stages, but not without some hesitation and uncertainty (see figure 4.4).

The Bohemian area of the Hercynian facies has remained an important one because of the relatively undeformed and continuous sequence of rocks and the abundance of their fossils.

Erben (1962, 1964) reviewed the distinctions between the Rhenish and Hercynian rocks and their fossils (see figures 4.5 and 4.6) and preferred the term *Magnafacies*, which signifies a facies group of several similar lithofacies and biofacies in chronological and geographical association. The terms have subsequently been used in the following way:

Rhenish   neritic faunas and near-shore shallow water, locally immature, clastic sediments and carbonates (including reefs).
(Associations of shelf facies.)

Hercynian   mostly pelagic or nektonic faunas in offshore deeper-water fine-grained sediments, pelagic and bathyal sediments.
(Associations of basin facies.)

They are no longer used to indicate separate faunal provinces despite the striking dissimilarities of the biota. They merge in many places, and elements of the one may be found in areas where those of the other magnafacies dominate. Shirley (1963) recognised the broad spread of the two magnafacies in Europe and throughout the proto-Tethyan (Rheic Ocean) belt as far as Australia. Work by Czechoslovakian geologists (see figure 4.7) has now provided very well-documented biostratigraphic units (zones) and chronostratigraphic units (stages) which may be identified far beyond the immediate type areas (Chlupáč, 1968, 1976).

One of the most important developments from the study of the Hercynian facies has been the recognition of the ammonoid succession. It has proved to be recognisable throughout much of the world.

In the last three decades stratigraphic Devonian studies in all parts of the world have sought to establish correlation with the classic sections in the Rhineland and Bohemia. Ammonoids, conodonts and tentaculites, essentially pelagic forms, have

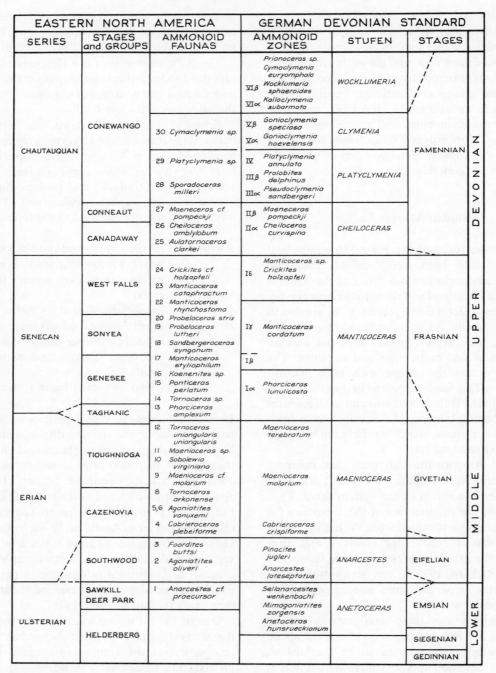

EASTERN NORTH AMERICA			GERMAN DEVONIAN STANDARD			
SERIES	STAGES and GROUPS	AMMONOID FAUNAS	AMMONOID ZONES	STUFEN	STAGES	
CHAUTAUQUAN	CONEWANGO		VIβ *Prionoceras sp.* *Cymaclymenia euryomphala* *Wocklumeria sphaeroides* VIα *Kalloclymenia subarmata*	WOCKLUMERIA	FAMENNIAN	DEVONIAN
		30 *Cymaclymenia sp.*	Vβ *Gonioclymenia speciosa* Vα *Gonioclymenia hoevelensis*	CLYMENIA		
		29 *Platyclymenia sp.*	IV *Platyclymenia annulata*	PLATYCLYMENIA		
		28 *Sporadoceras milleri*	IIIβ *Prolobites delphinus* IIIα *Pseudoclymenia sandbergeri*			
	CONNEAUT	27 *Maeneceras cf. pompeckji*	IIβ *Maeneceras pompeckji*	CHEILOCERAS		
	CANADAWAY	26 *Cheiloceras amblylobum* 25 *Aulatornoceras clarkei*	IIα *Cheiloceras curvispina*			
SENECAN	WEST FALLS	24 *Crickites cf. holzapfeli* 23 *Manticoceras cataphractum* 22 *Manticoceras rhynchostoma*	Iδ *Manticoceras sp.* *Crickites holzapfeli*	MANTICOCERAS	FRASNIAN	UPPER
	SONYEA	20 *Probeloceras strix* 19 *Probeloceras lutheri* 18 *Sandbergeroceras syngonum*	Iγ *Manticoceras cordatum*			
	GENESEE	17 *Manticoceras styliophilum* 16 *Koenenites sp.* 15 *Ponticeras perlatum* 14 *Tornoceras sp.*	Iβ Iα *Pharciceras lunulicosta*			
	TAGHANIC	13 *Pharciceras amplexum*				
ERIAN	TIOUGHNIOGA	12 *Tornoceras uniangularis uniangularis* 11 *Maenioceras sp.* 10 *Sobolewia virginiana* 9 *Maenioceras cf. molarium*	*Maenioceras terebratum* *Maenioceras molarium*	MAENIOCERAS	GIVETIAN	MIDDLE
	CAZENOVIA	8 *Tornoceras arkonense* 5,6 *Agoniatites vanuxemi* 4 *Cabrieroceras plebeiforme*	*Cabrieroceras crispiforme*			
	SOUTHWOOD	3 *Foordites buttsi* 2 *Agoniatites oliveri*	*Pinacites jugleri* *Anarcestes lateseptatus*	ANARCESTES	EIFELIAN	
ULSTERIAN	SAWKILL DEER PARK	1 *Anarcestes cf. praecursor*	*Sellanarcestes wenkenbachi* *Mimagoniatites zorgensis* *Anetoceras hunsrueckianum*	ANETOCERAS	EMSIAN	LOWER
	HELDERBERG				SIEGENIAN	
					GEDINNIAN	

Figure 4.4. Eastern North American and German ammonoid successions, stages and series compared (after House, 1979). Sequences that yield many of these zonal forms are also known in Morocco, Spain, the U.S.S.R. and China. The American stages are based upon the remarkable succession seen in New York State and were employed as far back as the middle part of last century.

RHENISH MAGNAFACIES	HERCYNIAN MAGNAFACIES
LITHOFACIES CHARACTERS	
Sediments 'impure': conglomerates, greywackes, sandstones, quartzites, quartzitic shales etc.	Sediments 'pure': mostly pure limestones and argillaceous shales
Calcium carbonate content low or absent	Content high, as a rule
BIOFACIES CHARACTERS	
CORALS	
Solitary corals almost absent	Very common and typical
Tabulates rare or absent	Very common and typical (*Favosites*, *Chaetetes*, *Heliolites*)
Pleurodictyum typical	*Pleurodictyum* absent
BRACHIOPODS	
Strongly ribbed forms predominant	Considerably less common
Smooth and rounded shells rare	Typical (pentamerids, meristids etc.)
Ribbed spirifers with long 'wings' are typical and very frequent	Smooth and rounded spirifers are typical and frequent
BIVALVIA	
Cardioconchs nearly absent	Present and typical (*Panenka*, *Kralowna*, *Buchiola* etc.)
Conocardium rare	Sometimes very common
Hercynella absent	Present and typical
Typical: *Grammysia*, *Phthonia*, *Leptodomus*, *Sphenotus*, *Limoptera*, *Dechenia*, *Modiomorpha* etc.	All absent or very rare
TENTACULITIDA	
Tentaculites predominant	Nearly absent
All absent or rare	*Nowakia*, *Guerichina* and *Styliolina* frequent and typical
CEPHALOPODA	
Nautiloidea rare	Very frequent and numerous
Arthrophyllum present	Absent
Bactritida always absent	Present and typical
Goniatites as rare exceptions only	Present and typical
ARTHROPODA	
Ostracods generally sculptured	Nearly always smooth
Trilobites mostly poor in number of genera and individuals	Rich in number of genera, species and individuals
Typical: *Asteropyge* and related genera. *Homalonotus* and related genera	All absent
All absent	Typical: Proetidae, Otarionidae, Harpidae, Scutelluidae, Odontopleuracea, Lichacea, Calymenidae, Cheiruridae, *Odontochile*, Phacopidella

Figure 4.5. The distinguishing features of the Rhenish and Hercynian facies are listed. The faunal characteristics of each are dependent upon the environments that are represented by sandy clastic facies in the Rhenish and argillaceous and calcareous facies in the Hercynian. These differences have long been known and they clearly represent different palaeoenvironments rather than different faunal provinces (see Erben, 1964).

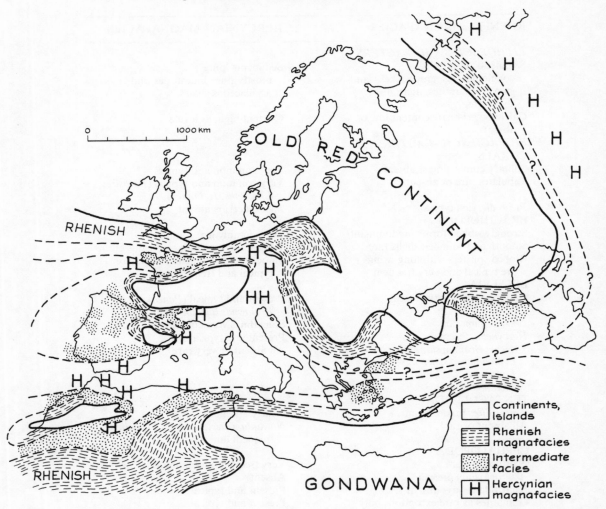

Figure 4.6. A suggested distribution of Rhenish and Hercynian facies in late Early Devonian time (after Erben, 1964). The Hercynian facies appear generally furthest from the margins of the land masses but in some areas the Rhenish and intermediate facies seem to be greatly reduced or missing from this lateral arrangement. Possibly this reflects local sedimentary 'starvation'.

been sought and studied to this end. Of these the conodonts have been the most spectacularly successfully employed and W. Ziegler's work on a conodont zonation of almost thirty zones stands as the probable means by which most correlation problems within the system will be resolved.

Correlation between the marine and the non-marine facies, however, is another matter, as few of the vertebrates stratigraphically useful in the Old Red Sandstone occur in marine strata. Here the use of palynomorphs in stratigraphy is increasing, and since these fossils occur in both marine and continental facies, it gives hope for better correlation than has been possible hitherto.

Between the 1870s and the 1960s was gathered the bulk of the stratigraphic detail upon which the current division of the system has been based. The distribution of rock types, the details of facies and the ranges of characteristic fossils became known not only in the extensive Rhenish Massif section but in many other parts of the world.

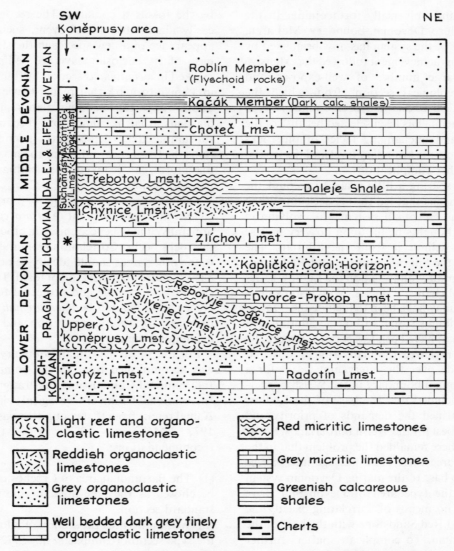

Figure 4.7. The Hercynian magnafacies of the Devonian is splendidly developed in Czechoslovakia where the rock units and stages near Prague constitute the Hercynian type section (after Chlupáč *et al.*, 1977). Biostratigraphic indexes such as tentaculitids, goniatites and graptolites enable a fair degree of correlation to be made with the Rhenish magnafacies.

The European Standard

Since about 1950 micropalaeontological techniques have been widely employed in enlarging and improving our knowledge of Devonian biostratigraphy. Spores, conodonts and ostracods, commonly obtained from rocks that were lacking in (determinable) macrofossils, have greatly aided the correlation of Devonian strata between one place and another. This has led to the state where in addition to macrofossil evidence the microfossils allow the drawing up of precise and internationally agreed definitions of the boundaries of the Devonian series and stages, and evaluation of the many different zones. It is not to be imagined, however, that such international agreement was immediately and universally forthcoming. In the case of the Siluro–Devonian boundary (McLaren, 1977) the base of the system has now been fixed by agreement. This achievement followed the extensive study of a large number of well-exposed and continuous geological sections from which the ranges of critical fossils had been found. A type section in Bohemia has been selected because of the fossiliferous and unchanging facies of the rocks above and below the boundary level and the lack of tectonic or other complications.

It is worth sparing a little space to recount the events and reasons that led to this decision. When the idea of the Devonian system was evolving in Murchison's thoughts he hit upon no very precise boundary to delimit it from the Silurian. The passage from the Silurian into the Old Red Sandstone of Shropshire follows upon the Ludlow series with a gradual change of facies. Murchison's original intention may have been to include the 'Downton Castle Building Stone' within the Silurian, but later he seems to have been less certain. Eventually White (1950) reviewed the many suggestions that had been made over the intervening years for a Silurian–Devonian boundary in the Welsh borderlands. He concluded that the base of the Ludlow Bone Bed satisfied the demands of priority and practicability best for the Welsh borderlands. So far so good, but there remained the problem that in the Silurian-type area there is no passage into *marine* Devonian. No base to the marine Devonian system is known in the type area of Devon and there seemed to be no means of correlating exactly the base of the Old Red Sandstone with the base of the marine Devonian. To complicate matters further Shirley (1938) and Boucot (1960) showed that a brachiopod fauna of pre-Gedinnian and post-Ludlovian age exists in Germany. This emphasised that the equivalent strata in the Welsh borderlands could not be readily recognised and defined. A new standard reference section was needed.

Meanwhile the philosophical basis for fixing geological time/stratigraphic boundaries had been energetically discussed and the concept of the 'golden spike' introduced. This holds that as geological time passes sediment is deposited (enclosing fossils) *without interruption* somewhere on earth. Each lamina of sediment represents an instant of geological time; each may be recognisable by the fossils it contains. The recognition of an easily identifiable faunal event, such as the first appearance of a new species in the strata, and the base of its enclosing layer of sediment can be marked in a *continuous* section by (literally) driving in a spike to mark that plane. However, as noted above (page 13), bedding planes may represent diastems (that is, time 'lost' because it is not substantiated by deposition). In continuous sections the fossil record, the unbroken evolution of a fossil form, can be taken as indicating an unbroken passage of time.

Ideally, then, a time plane or horizon may be *defined* at a specific point (a *stratotype*) and then, hopefully, it may be recognised in other areas where the rocks occur. Correlation from the type section to the other areas should ideally be possible on the basis of more than one kind of fossil. The stratotype should be accessible and protected for future reference and it should be *internationally accepted* as the basis for future correlation and reference in defining a boundary.

For the Siluro–Devonian boundary the business was accomplished by a special committee of the International Commission on Stratigraphy for the International Union of Geological Sciences in 1960. A final report from that committee was presented in 1977 (McLaren, 1977) with three prime recommendations:

(1) The Barrandian area of Czechoslovakia should be chosen as the type area for the international standard section.
(2) Within this area the section of Klonk should be the type section (Chlupáč *et al.*, 1972).
(3) The boundary horizon should be drawn within the 7–10 cm thick bed No. 20 immediately below the sudden and abundant occurrence of *Monograptus uniformis* and *M. uniformis augustidens* in the upper part of that bed.

Not everyone was happy with this recommendation but it has been ratified and upheld subsequently and the means of correlation between the marine sequence at Klonk and the non-marine sequences elsewhere are being sought. No one could deny that an energetic search had been made for the best section to use for the golden spike. Close examinations of sections and their zonal fossils in

Europe, North Africa, North America and the U.S.S.R. had been carried out by the committee and submissions of data from other areas had also been considered. The section at Klonk (see page 12) seemed to fulfil all the requirements—it is accessible, well preserved and unbroken, understood in terms of physical stratigraphy and biostratigraphy and involves no facies changes; comparable sections are also available nearby.

So here we have a boundary that is defined as the base of a zone and is the boundary between the Hercynian stages Pridoli and Lochkov. Not only is this zone marked by the first appearance of the graptolites named above but within it are also the first specimens of the trilobite *Warburgella rugulosa rugulosa*, which is useful because this fossil occurs in limestones where the graptolites do not. Similarly the conodonts *Icriodus woschmidti* and *Spathognathodus steinhornensis eosteinhornensis* are good marker species for the zone.

As Walmsley (1974) points out in a very useful account of these deliberations, the consequences of the recommendations being accepted leave Britain without a stratotype for a boundary of some importance and between two systems based upon British geology. Despite the search, no graptolites have been found above the *M. leintwardinensis* zone here, nor are they likely to be. The question of how the Old Red Sandstone facies is to be biostratigraphically correlated with the *M. uniformis* horizon and the Pridoli stage is discussed below (page 59).

The Lower Devonian

Between 1849 and 1900 came the publication of several accounts of the succession in the Rhineland and the introduction of stratigraphical names that are still used. The Belgian, Dorlodot's, classification (see figure 4.1) was eventually adopted by almost all concerned, even though the Emsian Ems-Stufen was not formally accepted in Germany till the 1950s. Even so, the three Lower Devonian stages were defined largely in lithostratigraphic terms and only later were (perhaps inadequate) biostratigraphic criteria applied. Disappointingly, the most important fossil groups (conodonts, graptolites) for correlation are largely missing from the Gedinnian and Siegenian type areas. (Some idea of the details

of the type sections for the stages and of the varying usage of stage names can be seen under the stage names in the *Lexique Stratigraphique International* (Pruvost, 1956).) In the Eifel area similar work was carried out and the ranges of many fossils, their dependence upon facies and their phylogenetic changes have been noted. Meanwhile in the vicinity of Prague the Zlichovian, Pragian, Dalejan and Lochkovian stages were designated for the Lower Devonian there. They do contain conodonts and graptolites and can be correlated widely. A possible combination of Rhenish and Bohemian styles may eventually be suggested for the Lower Devonian standard. These studies will continue.

The Middle Devonian

Perhaps because it is so widespread and fossiliferous, laterally variable and appears in many separate outcrops, the series referred to as Middle Devonian has not always been drawn between the same stratigraphic levels. In the classic Belgian and Rhineland areas two different lower boundaries have been used and different names have been employed for the lower of the two stages that are included in the series. French-speaking geologists have preferred the name Couvinian, while others have used the name Eifelian.

Of the two terms, Eifelian seems to have been first in use although it was subsequently modified several times. Couvinian was designated in 1862 some fourteen years later than its rival but has not been so imprecisely used by subsequent workers (Pruvost, 1956). Even now agreement has only recently been reached on the precise correlation of the Belgian and German sections. The most generally accepted level that has been proposed for the base of the series is at the base of the *Polygnathus costatus partitus* and stratotypes for the boundary have been selected at Prüm in West Germany and at Holyň Prastov in Czechoslovakia.

The Upper Devonian

Two stages, Frasnian and Famennian, are customarily used for the subdivision of the Upper Devonian. The Frasnian, the Senecan (North

America), the Adorfian and the *Manticoceras*-Stufe are all names that have been applied to an early Upper Devonian stage. Certainly they cover approximately the same span of time but precisely how synchronous they are remains uncertain. Belgian geologists have favoured the inclusion of the Assise de Fromelennes in the Frasnian although the base of the Assise de Frasnes has been advocated. The German ammonoid succession has a useful datum in the sharp faunal break between the zones of *Maenioceras terebratum* and *Pharciceras lunulicosta* and although the Assise de Fromelennes has yielded no ammonoids its base must lie close to the break just mentioned. Detailed work on the ranges of conodonts and ammonoids (House and Ziegler, 1977) supports the idea that a level close to the base of this Assise might be chosen as a precisely defined palaeontological boundary. They offered three possibilities:

1. base of the ammonoid zone   *lunulicosta*
2. base of the conodont zone   *hermanni-cristatus*
3. base of the Lower of the
   conodont zones of   *P. asymmetricus*

Of these the last seems to be favoured most and a stratotype in pelagic facies, possibly in France or Morocco, may soon be chosen.

The Famennian seems now to be relatively uncontroversial in that the German successions with ammonoids and conodonts can be correlated with a Belgian succession. The top of the stage and hence of the Devonian system itself has been drawn at the base of the *Gattendorfia* beds, as advocated by an international meeting to fix the base of the Carboniferous system on the best fossil evidence available (Pruvost, 1956).

What emerges from this seemingly pernickety arguing about the limits of the Devonian system and its subdivisions is important. There is a need to reconcile the different successions in Western Europe and elsewhere on the basis of a common palaeontological standard. Facies differences and facies faunas have made it difficult but the recognition that conodonts have an extraordinary stratigraphic sensitivity (that is, they evolved rapidly) and have an almost universal distribution has suggested a means of achieving this. Though by no means so widespread, the ammonoids offer a sim-

ilarly refined succession. Between these two fossil groups correlation is possible and the ammonoid chronology can be applied indirectly to facies where the ammonoids themselves do not occur. Further testing of this model in distant parts of the world has been actively prosecuted in recent years and international debate should reach agreement on the results soon.

The U.S.S.R.

The enormous territories of the U.S.S.R. include Devonian rocks on the Russian and Siberian Platforms and within the orogenic belts of the Urals, Caucasus, Tienshan, Kazakhstan, Altai-Sayan Taimyr, Verkhoyansk-Chukotsk and Mongolo-Okhotsk (Rzhonsnitskaya, 1968; Nalivkin, 1973a, 1973b). There are extensive Old Red Sandstone types of deposit as well as marine formations and volcanic rocks on an enormous scale in the eastern Urals, Kazakhstan, Altai and the Omolon block (see figure 4.8). Palaeontologically, two distinct marine realms, Atlantic and Pacific, are recognised with several 'provinces' in each.

Soviet stratigraphic nomenclature is rather different from that in Western Europe in the use of such words as 'epoch', 'suite', 'tolshcha' and 'horizon'. *Epoch* may be used where *age*, the time equivalent of a stage rather than a series, is implied. A *suite* commonly approximates to a stage or larger subdivision which may actually include representatives of more than one system. A *tolshcha* corresponds to a small or localised formation, but may be used for a system. The term 'horizon' as a biostratigraphic unit of somewhat uncertain rank is frequently used and may correspond to a substage or zone or merely indicate 'beds'.

The use of these seemingly rather loose terms does not, however, appear to cause the confusion one might expect in translation. A good review of Soviet stratigraphic practice is given in Nalivkin (1973a) and biostratigraphic progress has been reported by Sokolov and Rzhonsnitskaya (1982).

There is no shortage of good unbroken sections on which to base local stages but the Soviet geologists have always attempted to correlate immediately with the stages of Western Europe. The base of the system has locally been redrawn to

Figure 4.8 — Correlation of Devonian successions in the U.S.S.R.

Central Russian Platform Urals (Substages)	Western Slope of the Urals (Suites)	Eastern Slope of the Urals	Central Asia — Superhorizon / Beds	Central Asia — Superhorizon or Stage	Key conodont levels	Salair — Horizon	Salair — Superhorizon or Stage	European Standard — Super-horizon (=Series)	European Standard — Stage	European Standard — Series	European Standard — System
Upper Famennian; Lower Famennian	Dankovo-Lebedyanian; Yeletsian; Zadonskian	Murzakayevskian (20m); Makarovian (18m); Zalairian Series: Kiian Suite (300–400m); Bakeshevskian Suite (400–500m); Karantauian Suite (10–500m); Askynian (350m)	Pistamazarian	Famennian	*disparilis*	Podonino; Pestchorka; Solomino; Hlubokaya; Kurlyak; Teryokhino; Vassino; Izyly	Famennian	Upper (or Superhorizon C)	Frasnian Famennian	UPPER	DEVONIAN
Upper Frasnian; Middle Frasnian; Lower Frasnian	Voronezhian; Yevlanovian; Livenian; Rudkinian; Semilukian; Alatyrian; Yastrebovskian; Shchigrovian Superhorizon	Barmian (5m); Domanikian (35m); Mendymian (12m); Koltubanian Series (?0–100m); Pashiskian (0–25m); Sargayevian (12m)	Shirdagian	Frasnian	*L. asymmetricus*	Akaratchkino; Kerlegesh; Safonovo; Alichedat	Frasnian	(Upper, cont.)	(Frasnian)	UPPER	DEVONIAN
Givetian	Khvorostanian; Starooskolian; Vorobyevkian; Olkhovskian; Mosolovian	Tatlybayevskian Suite (25–300m); Askarovian Suite (150–200m); Aushkulian Suite (0–150m)	Zinzilban Beds; Novikhushk Beds (= Kim Beds); Dzhaus Beds; Obisafit Beds		*c. costatus*; *c. serotinus*; *c. patulus*	Mamontovo; Shanda; Belovo Salairka	Teleutian; Telengitian	Middle (or Superhorizon B)	Eifelian Givetian	MIDDLE	DEVONIAN
Eifelian	Morsovian; Ryazhskian; Novobasovian		Rushsayan		*dehiscens*	Maliy Batchat	Beltirian	Middle (or Superhorizon B)	Eifelian Givetian	MIDDLE	DEVONIAN
(Lower Devonian)	Tanalykian Suite (0–75m); Bugulygyrian Suite (50–100m); Irendykian Suite (500–1000m)	Kalmykovian Suite (350–400m); Kusimovskian Suite (30–50m); Khasanian Suite (25–50m)	Madmohiah; Bursykhirmanian; Kushnovian		*sulcatus*; *pesavis*; *woschmidti*	Tom Tchumysh; Peetz; Krekov	Kaibalian	Lower (or Super-horizon A)		LOWER	DEVONIAN

Figure 4.8. The vast expanse of the U.S.S.R. reveals Devonian rocks in many parts; four of the most important and continuous successions are shown. They illustrate the different style of subdivisions that may be used by Soviet stratigraphers. The Central Asian and Salair regions have recently been very intensively studied and the Middle Devonian seems much reduced in thickness. Many Soviet stratigraphers favour the division of the Devonian into two series only—an idea that does not find favour elsewhere.

conform to the practice now advocated by the International Commission on Stratigraphy (McLaren, 1977).

Several terms taken from the classical western sequence of stages are, however, used rather differently in the U.S.S.R. Thus 'Gedinnian' has been used for beds that correspond to those of the Upper Gedinnian and Siegenian in the west. The 'Coblenzian' stage occupies the remainder of the Lower Devonian, extending up to the base of beds with *Favosites regularissimus* or *Uncinulus parallelepipedus*. The Middle Devonian thus seems to begin with beds that are equivalent to part of the Upper Emsian in Western Europe and the 'Eifelian' stage extends up to the base of the Piarnu horizon of the Russian Platform. The 'Givetian' continues thence up to a break in sedimentation above which 'Frasnian' is established.

'Frasnian' and 'Famennian' stages have lower boundaries that are defined by the base of the zones of *Koenites nalivkini—Cyrtospirifer murchisonianus* and *Cheiloceras–Crystospirifer archiaci* respectively. The base of the Carboniferous system is drawn at the base of the zone of *Wocklumeria*, as it commonly is elsewhere.

China

The Chinese Academy of Geological Sciences has since 1979 adopted a vigorous programme of stratigraphic research and has been actively promoting study of correlation of the Chinese regional stages and facies with those of the rest of the world. It is a country rich in Devonian rocks (Hou *et al.*, 1979; Yang *et al.*, 1981) and with both marine and continental formations in great variety (see figure 4.9). They make increasing use of conodonts, dacryoconarids and macrofossils for correlation and their stages are based upon good type sections

SERIES	STAGES	TYPE LOCALITIES	LITHOLOGY AND FOSSILS
UPPER	Hsikuangshanian (Famennian)	Xinghua County, Hunan Province	argillaceous rocks with vertebrates and plants in the upper part and brachiopods below
	Shetienchiaoan (Frasnian)	Shaotung County, central Hunan Province	limestones and siltstones in the corals; brachiopods and ammonoids
MIDDLE	Tungkanglingian (Givetian)	Xiangzhou County, eastern Guangxi Province	clastics and limestones with brachiopods and coral faunas
	Napiaonian (Eifelian) (Dalejian)	Nandan County, northern Guangxi Province	black carbonaceous mudstone with ammonoids, tentaculites, trilobites and brachiopods / parastratotype of limestones and mudstones with brachiopods
LOWER	Tangdingian (Zlichovian)	Nandan County, northern Guangxi Province	dark mudstones with ammonoids and tentaculites and a parastratotype of limestone with rich brachiopod faunas
	Yukiangian (Pragian)	Hengxian County, Guangxi Province	siltstones and limestones: brachiopods, corals, trilobites, conodonts
	Nakaolingian (Lochkovian)	Hengxian County, Guangxi Province	mudstones and limestones: with brachiopods, corals, spores and conodonts
	Lianhuashanian (Lochkovian)	Hengxian County, Guangxi Province	largely continental clastics; with vertebrates

Figure 4.9. The Devonian system in China is widespread and varied from the north to the south. Several different 'types' (magnafacies) are known and correlation with the European stages has been successfully attempted. However a system of local stages has been set up and recognised in the different facies developments in China. Here correlation with the Hercynian stages of the European Middle and Lower Devonian is suggested (Yang *et al.*, 1981).

in the appropriate facies. Correlation of Chinese stages with those based upon European type sections is still uncertain in detail. The faunas clearly belong to the Old World Realm, and several of the critical horizons for the delimitation of European stages (and series) can be identified in China (Wang *et al.*, 1981).

Three main domains of Devonian rocks occur: a 'geosynclinal' belt of thick clastics that lies north of the Inshan-Tianshan mountains, a middle domain between the Inshan-Tianshan and the Qinling-Kunlun, and a very big southern domain south from the Qinling-Kunlun as far as the Vietnamese border.

Continental Facies

Biostratigraphic correlation is difficult as there are so few fossils that are common to both the continental and the marine facies. A natural consequence of this has been that where thick continental successions occur local zones (and stages) or horizons based upon vertebrates have been proposed. In Britain the Lower Devonian allows a subdivision of that kind, but a full sequence of zones or stages for the continental facies still does not exist. In Canada, the U.S.S.R. and China various groups of vertebrates have been pressed into service for a faunal subdivision of the continental successions there. It would be a temporary expediency if one of these successions were to be adopted as a type, since the ultimate aim in correlation is to have a single international and intercontinental chronostratigraphic time-scale. Nevertheless, the local stages proposed in Britain and elsewhere are useful in present discussions on correlation between continental basins. International correlation should soon be possible.

Stages of the British Continental Facies

The Old Red Sandstone of the British Isles does not easily lend itself to biostratigraphic subdivision. Early recognition of Lower, Middle and Upper Old Red Sandstone series in Scotland was based more upon structural relationships and lithologies than on palaeontological criteria. The lack of both stratigraphic continuity between continental and marine facies and fossils common to both facies in the west of England inhibited correlation there. Discoveries of vertebrates and plant remains in the intercalated marine and continental rocks of Europe were helpful in confirming stratigraphic ranges of some of the Old Red Sandstone vertebrates but not until the years between the two world wars were serious efforts made to erect in Britain a scheme of zones and stages for the Old Red Sandstone. The pioneer efforts of White (1950), Ball *et al.* (1961) and others has resulted in a biostratigraphic subdivision of these formations in Wales and the Welsh borderlands. Gross (1950) reviewed the geographical range of Lower, Middle and Upper Devonian vertebrate faunas.

Vertebrate faunas in mainland Europe, Spitsbergen and Canada (see figure 4.10) have been found to show enough varying but generally convincing similarity to permit biostratigraphic correlation (Dineley and Loeffler, 1974; Westoll, 1979). It should be emphasised that the collections of fossils upon which these biostratigraphic units are based are from scattered localities rather than from continuous sections and that type sections in the strict sense comparable to those in the marine facies have so far not been designated. In Scotland also formal biostratigraphic subdivision has not progressed beyond this level. In Canada, however, a continuous section from Silurian into Dittonian rocks with marine bands containing conodonts and also strata containing a phylogenetic sequence of pre-*leathensis* pteraspids has been found (Dineley and Elliott, 1983). It is dismaying that this section so far has yielded no spores by means of which correlation with known sections elsewhere could be made. Spores remain the best means of correlation between the continental and the marine facies (see page 81; also McGregor, 1979), but ostracods may locally be useful in this respect.

Zones based upon vertebrates in the continental successions of China and the U.S.S.R. have not yet been designated even though the ranges of vertebrate faunas and individual taxa are increasingly being measured with welcome accuracy.

With the adoption of a new base for the Devonian system the need to recognise its equivalent in the Old Red Sandstone of the Welsh borderlands brought along a number of suggestions (Holland

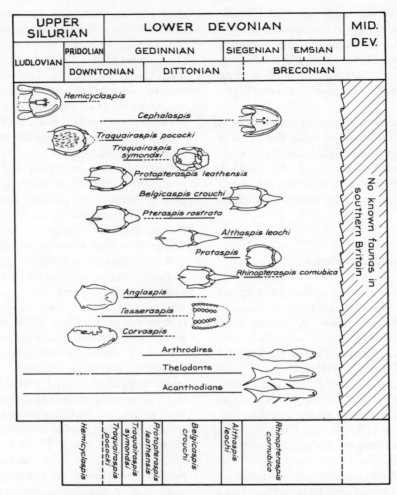

UPPER SILURIAN		LOWER DEVONIAN			MID. DEV.
	PRIDOLIAN	GEDINNIAN	SIEGENIAN	EMSIAN	
LUDLOVIAN	DOWNTONIAN	DITTONIAN		BRECONIAN	

Figure 4.10. The ranges of early vertebrates, here largely the agnatha, are of use in the biostratigraphy of the Lower Devonian (Old Red Sandstone) of Great Britain, Western Europe and eastern Canada.

and Richardson, 1977). The Downtonian stage seemed to be undeniably Silurian on the basis of new correlations with Europe. Martinsson (1967, 1977) is able to show the position of the Ludlow Bone Bed, the base of the Downtonian, in terms of the Baltic ostracod succession and hence achieves something of a link between the marine and non-marine facies. For the Downtonian/Dittonian boundary Allen and Tarlo (1963) had proposed that an appropriate level should be where the major palaeontological break in the region occurs. This seems to be, however, a facies-controlled boundary to some extent—where brackish intertidal sedi-

ments give way to fluviatile beds—which is very difficult to map. Later Tarlo (1965) thought that this boundary could also stand for the Siluro–Devonian boundary as well but Westoll *et al.* (1971) pointed out that this facies change (from Downtonian to Dittonian) must be diachronous. To define the top of the Downtonian by defining the base of the Dittonian at the base of the zone of *Pteraspis (Protopteraspis) leathensis* is in theory permissible but no highly fossiliferous stratotype for the boundary has been found. A good section in Shropshire (Holland and Richardson, 1977) with limited fossils is nevertheless useful as a temporary compromise. S. Turner (1971) has suggested that thelodonts may provide better correlation with Old Red Sandstone formations in Europe and North America and that the advent of the *Turinia pagei* fauna may be a suitable datum level for the Downtonian/Dittonian boundary. The level of the base of the *M. uniformis* zone may be close to this but so far proof is not available.

The exact status of the Breconian stage remains far from satisfactory. Proper biostratigraphic definition is needed but fossils are all too few. The pteraspid genera *Althaspis* and *Rhinopteraspis* may prove useful eventually.

Middle and Upper Devonian vertebrate biostratigraphy, as revealed in Scotland, has recourse to faunal assemblages rather than to strictly defined zones (see page 97).

Radiometric Ages

No matter how precisely the limits of a geological system may be defined on the basis of fossils, the result gives no clue as to the age in years of the rocks in question. To provide this essentially quantitative information radiometric age determination is needed, but this is fraught with difficulties. Methods of mineral and whole-rock analyses have improved greatly in recent years but anomalous results and revised estimates are discussed in the literature each year. Four principal criteria are required for accurate age determinations: the stratigraphic age of the samples must be known as accurately as possible; the radiometric data must be precise and unambiguous; samples of the minerals used should be described so that the origins of discrepancies might be identified; and the geological history of the parent formation should be known. Volcanics from Europe, North America and Australia have been particularly useful in providing potassium–argon, rubidium–strontium, uranium–lead and thorium–lead ratios upon which Devonian ages have been calculated.

Potassium is a very abundant element and about 0.4 per cent of all potassium is ^{40}K, the isotope which produces argon-40. The isotope occurs in micas, potash feldspars and hornblende and has a half-life of 1300 million years. The geological history of the parent rock is significant since deep burial of the nucleide-containing minerals is known to lead to the loss of the daughter element argon-40. Metamorphism and weathering may also, and perhaps more obviously, affect the retention of radioactive elements. Methods of analysis vary but enough data have been accumulated about the decay rates and other necessary physical parameters involved for the I.U.G.S. Subcommission on Geochronology (Steiger and Jäger, 1977) to recommend those that are more trustworthy. By means of radiometric 'whole-rock' analyses of igneous bodies minimum ages for their emplacement and crystallisation may be suggested with a degree of confidence. Many ages have to be given in the form of, say, 100 ± 10 million years, which means that successive analyses for the same rocks are likely to give ages within a range of 20 million years. Repeated determinations would give figures within 10 more or less than the mean 100.

Radioactive potassium is also found in the authigenic mineral glauconite. Devonian greensands are so far unknown but glauconites of this age would be regarded rather dubiously on account of the extreme susceptibility of argon to escape from the mineral structure. A consistent loss of argon is such that the age of most Cretaceous glauconites is about 5 per cent too young and Palaeozoic examples 10–20 per cent too young.

The information available to Friend and House (1964), Lambert (1971), and Van Eysinga (1975) suggested the following dates (in million years) for the Devonian period:

End of the Devonian	345
Upper/Middle Devonian boundary	358
Middle/Lower Devonian boundary	370
Beginning of the Devonian	395

Since then analytical techniques have advanced and some of the critical points in the calibration table of stratigraphical ages and radiometric values are now regarded as unreliable. Practice in field and laboratory has not everywhere been the same and different constants have been used on occasion. The number of reliable age determinations for Devonian materials that give useful points on a mid-Palaeozoic time-scale is still very small. Since these are rather widely scattered through the stratigraphic column, a plot of radiometric age against some quantity that is proportional to the actual time duration of the stratigraphic stages is needed. In the past maximum stratigraphical thicknesses were used but not all of these referred to the same kind of depositional environment and hence could not be thought of as reliable. Boucot (1975) discarded this approach and, being a palaeontologist, preferred the average rates of animal evolution as a guide to stage length. This, too, is liable to be a somewhat uncertain criterion, but may be acceptable enough where the rates of several contemporaneous fossil groups can be used. Gale *et al.* (1979) adopted it for a new plot of stage length against radiometric age. It gave a good least-squares regression line that indicates a minor revision to the previous figures but McKerrow *et al.* (1980) made a somewhat different interpretation of data from acid volcanics and fission-tracks. Odin (1982) offers yet other slightly different figures but there is clearly now a broad and reasonable consensus:

Boundaries

	S/D	LD/MD	MD/UD	D/C
Gale *et al.*	394	384	367	345
McKerrow *et al.*	411	390	377	360
Odin	395–410	377–393	370–380	350–365

Thus the first two papers suggest a length of about 40 million years for the period, rather than the 50 or 45 million years given by previous authors; Odin's figures invoke a greater length and his overall assessments of Phanerozoic dates have aroused some disagreement.

From what has now been said it may be understood that the necessary expediencies of 140 years ago are being replaced by practices that are as precise and rigorous as we can make them. It has always been the aim of stratigraphers to make correlation precise and to develop the means of achieving this throughout the world. As far as the Devonian system is concerned, the process is yet to be completed but the means seem to be within our grasp.

Age dating by radiometric means is a very different discipline but one necessarily taken in tandem with biostratigraphic correlation to provide aspects of proportion and perspective in historical geology. Devonian 'dates' are all too few and far between but give some help in understanding the dimensions of this system within the very broad confines of the Palaeozoic era.

5

Biostratigraphy

The basic principle that strata with the same palaeontological content are contemporary enables us to correlate strata from one locality to another and to construct a form of fossil calendar by which to denote spans of geological time. There are of course geographical constraints upon the spatial ranges of organisms but the first appearance of a species in rocks of a given area may represent a (local) evolutionary event, or the shifting of a favourable environment or the removal of a previous barrier to migration. Identifying which of these alternatives occurred is commonly difficult. In any local section it is usual to be able to distinguish simple *range zones* for the different fossils that are present. The local range zone may represent the total range of the index forms (biozone) or only part of it (teilzone). *Faunizones* or *florizones*, erected for many successions, are based upon faunal or floral *assemblages*, hence they are assemblage zones, while *epiboles* or *acme zones* are the beds in which the greatest frequency of named species occur. Biostratigraphy and inter-regional correlation is concerned with finding the best means of establishing unbroken faunal and/or floral successions, preferably on the basis of proven phylogenetic lineages. Many type sections are now being re-examined so that the validity of the zones originally erected within them may be tested over wide areas.

Ideally the good stratigraphic index or key fossil should be virtually independent of facies. In fact most fossils are restricted to particular facies and regions. Facies fossils are those that are most closely proscribed or restricted by ecological or facies conditions in the *biotopes* in which the organisms lived. Such conditions may not be limited in time or space and may give rise to *ecozones* in which the particular conditions that determine the fauna have

a wide extent or long history. Once a faunizone is terminated by the extinction of its key species the zone can never re-appear in the succession.

For most practical purposes stratigraphers regard the first appearance of a fossil in the Phanerozoic record as being synchronous wherever it is found. The first appearance of a new species or zonal fossil (a morphotype) is clearly the best criterion for defining the lower boundary of a zone. Defining zones by reference to several taxa makes for complications. It follows, too, that the lower boundary of one zone is the upper boundary of the zone beneath it. Ideally there should be no gaps between zones and no overlaps either (see Murphy, 1977). The unbroken stratigraphic sequence is taken to represent an unbroken span of geological time and the zone defined by the first appearance of one species and its later replacement by another is thus a true *time-stratigraphic* (*chronostratigraphic*) unit.

The relative values of different fossil groups for biostratigraphic purposes have been known for many years but the particular merits of relatively recently studied forms such as conodonts, acritarchs, spores and dacryoconarids seem to increase as work proceeds. Biozones of different fossil groups can overlap geographically and stratigraphically. The differing rates of evolution between groups rarely seem to allow different suites of biozones in a succession to share boundaries. The overlap of zones from one facies to another is also more infrequent than we would wish. Thus, eventually, spore floras may be recovered from both Old Red Sandstone and marine Devonian to allow these two very different rock assemblages to be correlated. Where the full stratigraphic span of a series can be recorded in terms of the zones of a particular fossil group it may be said to constitute an *ortho-chronology*. It becomes the 'calendar' or standard against which other schemes and time-scales (*parachronologies*) may be tested or compared. Selecting zonal boundaries in stratigraphic sections may be guided by various principles but the *first* appearance of a zonal form is not to be discarded as a preferred criterion.

It is not possible here to do justice to the living world of 360–410 m.y.b.p. but a brief survey of the animal and plant kingdoms of the Devonian period is useful before we examine the biostratigraphic use of some of the fossils of the time. The difficulties of understanding the biotas of the past are immense. Most of the organisms that were there have left no trace: in most instances we are presented with a few fossilised hard parts — a very biased sample. Accepting that what is actually recorded is but a small fraction of the Devonian flora and fauna, it is interesting to see what the record includes and what general observations may be made.

The Devonian macroflora is important in that it presents so much that is new in plant morphology, despite the fact that the fossils are usually preserved as fragments. H. P. Banks (1980) at Cornell University has shown that Devonian plant megafossils, albeit fragmentary, do provide useful assemblage zones. These matters are referred to in more detail in chapter 6.

The current debate about the apparently increasing diversity of life throughout geological time is not without relevance. The long-held concept that the number of (marine invertebrate) species has varied throughout the Phanerozoic (see figure 5.1), reaching its maximum in the Recent, has been challenged. An alternative view is that the habitable environment was soon filled by the maximum diversity of living forms that it could support and that this number has changed little since then. D. M. Raup of the University of Rochester has held that species diversity is strongly correlated to volume of rock available and perhaps diversity among marine invertebrates has been at saturation point or equilibrium has been achieved throughout the Phanerozoic (Raup, 1976b).

In an attempt to estimate the numbers of fossil species recorded Raup (1976a) made a survey of the entries (for nineteen taxonomic groups) in the *Zoological Record*, listing their numbers system by

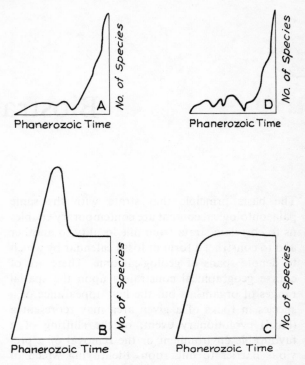

Figure 5.1. The diversity of marine shelf invertebrate species during Phanerozoic time has been differently postulated by several palaeontologists. Valentine (1970) showed an irregular curve with a strong post-Palaeozoic increase (A). His later (1973) model (D) showed even stronger changes in the curve. Raup's model (B) shows a Late Palaeozoic peak and subsequently a plateau of equilibrium (Raup, 1972); his model (C) lacked the strong peak of the earlier one (Raup, 1976b). All indicate that the latter half of the Palaeozoic was a time of increasing diversity, the Famennian extinctions notwithstanding.

system. Treatment of the data obtained shows 'a Palaeozoic high in the Devonian which is approximately four-tenths of the Cenozoic level' (Raup, 1976a).

House (1975b) records more than 1100 for the number of Middle Devonian marine invertebrate genera, the maximum for the era; the Upper Devonian follows with over 900, and of these a mere 280 survived into the Carboniferous. Family numbers reached about 330 in the Middle Devonian, 290 only surviving into the Late Devonian. It would, however, be unwise to read too much into these figures as species, genera and families are somewhat

Zoological Record 1900–1970
The Devonian biota: numbers of species recorded (after Raup, 1976a)

	Spp.		Spp.
Foraminifera	291	Cephalopoda	423
Radiolaria	79	Other Mollusca	19
Other Protozoa	193	Trilobites	1045
Porifera/Archaeocyathids	332	Crustacea	1609
Coelenterata	2868	Insecta	2
Bryozoa	350	Other Arthropoda	50
Gastropoda	314	Echinoidea	4
Bivalvia	532	Other Echinodermata	541
Brachiopoda	1726	Vermes and Trace fossils	761
		Hermichordata	10
		TOTAL	11 449

The *total* number of Devonian species Raup calculates as 12 139

subjective entities. It is also tempting but questionable to suggest that the greatest diversity of the Devonian fauna occurred at the moment of widest marine transgression and that numbers diminished thereafter. Differing conditions prevailed from one marine environment to another and, with rising sea level in Frasnian time, changes occurred in many of them. There followed an almost world-wide halt t reef building and a collapse of many carbonate communities.

The Devonian does show a peak in diversity but as Raup (1976b) points out, the picture is likely to be distorted by biases of preservation, collecting and so on. If the apparent increase from the Silurian to the Devonian is true (per million years) there are several possible explanations:

(1) The number of species co-existing in each habitat could have increased on a world-wide basis.
(2) The diversity of one habitat (reef?) could have increased on a world-wide basis.
(3) The total habitable area could have increased through changes in geography (perhaps affecting the areas of continental shelf) or through amelioration of climate.
(4) Provinciality could have increased as a result of changes such as given in (3) above.

As mentioned above, Valentine and Moores' (1970, 1972) plate-tectonic model (see figure 5.2)

and the correlation of its behaviour with apparent species diversity levels seeks to explain the increase. If, however, the apparent increase is only the result of bias in the fossil record, the most likely time-dependent factors to have influenced it are:

(a) **Area, thickness and volume of exposed sedimentary rocks.**
(b) Degree of diagenesis and metamorphism.
(c) Amount of exploration.

Non-time-dependent factors include:

(d) Errors in radiometric time-scale.
(e) Change in evolutionary turnover rate.
(f) Evolution of organisms with different preservability.
(g) **The sedimentary facies that dominate the period.**
(h) Monographic effects.

Devonian species diversity correlates with the area and volume of sedimentary rock available and some 270 species appear per million years during this period as against 180 species for the Silurian and 166 species for the Carboniferous per million years (Raup, 1976a). It was a time of high sea level and a large habitable area of shelf seas existed. Sepkoski (1976) suggested that larger areas provide the possibilities of a greater variety of habitats than the smaller areas. Species in larger areas have lower

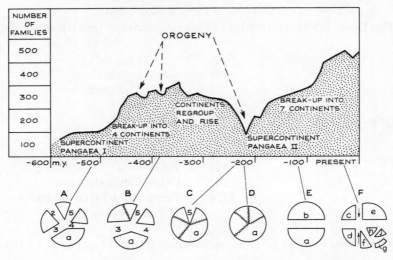

Figure 5.2. The behaviour of the major continental masses and the fluctuations in the numbers of well-skeletonised marine invertebrate animals were related in a simple plate-tectonic model by the Americans J. W. Valentine and E. M. Moores in the early-1970s (Valentine and Moores, 1972, 1974). The lower figures A–F suggest the relative configurations of the continents and oceans. The Devonian period was distinguished by three, possibly four, major continents which were to become widely flooded before the period ended.

rates of extinction and hence more species are eventually present. Environmental change may be said to prompt organic evolution (see page 16).

The conclusions to be reached from these discussions are broadly supportive to the idea that there was a real peak in the diversity (or species richness) of organisms present in the Middle Devonian shelf seas. There were many innovations especially in the evolution of reefs, arthropods, ammonoids, plants and vertebrates, to name but a few.

Devonian *Dramatis Personae*

The number of taxa that were present in Devonian times but which have left little or no recognisable trace of their existence must be many times as large as that of the fossilised remains. Many soft-bodied invertebrates living then have probably left no descendants. The record of non-marine (aquatic) organisms is equally tantalising in its obscurity. Figure 5.3 shows the varying contribution that each of the major invertebrate groups has made to the fossil record (Raup, 1976a).

Protista

The unicellular organisms have a poor record in the Devonian rocks but the taxa present indicate a wide range of types and modes of life. More than 700 species of marine acritarchs are known and many are ubiquitous. Radiolaria were widespread and the calcareous foraminifera appeared in the late Middle Devonian, perhaps in response to the niches that were offered by the spreading epicontinental seas of the time. Agglutinated foraminifera existed but have not been widely recognised. Hystrichospheres were abundant and the first volvocale Chlorophytes are known from the Upper Devonian. On the other hand, the (planktonic?) Chitinozoa became extinct.

Porifera

The simplest metazoa were present in a wide variety of marine environments, from the very shallow to the deep sea. They seem to have been only locally common, as in New York State, where the delicate skeletons of *Hydnoceras* and other genera occur.

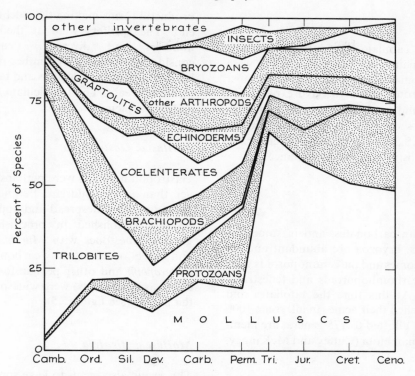

Figure 5.3. Variations in the proportions of the major invertebrate groups during Phanerozoic time (after Raup, 1976a) showing the brachiopods and coelenterates at their peak in the Devonian, together with a conspicuous increase in arthropods and other invertebrates. Trilobites show a reduced presence and molluscs overall show a small reduction.

Receptaculites, a problematic fossil but commonly regarded as a large calcareous sponge, is abundant in some carbonate facies in North America and is widespread elsewhere also.

Stromatoporoidea

This group has been given several kinds of status over the years and it has even been regarded as possibly being cyanophycean algae. The organisms were the dominant reef-builders and were common members of the carbonate community. Lower Devonian genera are few, but 45 per cent of all known species (over 450) are of Eifelian to Frasnian age. Highly endemic genera occur in the Famennian in different parts of the world.

Rugosa

Although rare in the Early Devonian, rugose corals enjoyed a remarkable increase in numbers during Middle and Late Devonian times, taking advantage of the wide shallow seas in warm-water latitudes. Isolated and colonial forms both increase in size and diversity and the colonial rugosa are important members of the reef communities. The isolate forms are found in many environments, including those associated with reefs. Oliver (1976, 1977, 1980) has described the palaeogeographical distribution of Devonian coral faunas—the relatively small Eastern North American Realm, the larger Old World Realm and the Malvinokaffric Realm.

Tabulata

The heyday of the Tabulata came during the Middle and Late Devonian when all the families were present and Devonian reef growth was at its most vigorous. This group seems to have experienced little endemism and is abundantly represented in most carbonate sequences. It has been closely studied in the U.S.S.R. where some species have been accorded biostratigraphic importance.

Bryozoa

In many Devonian carbonates, and in some fine clastic formations, bryozoa are abundant and diverse. The trepostomes and the ceramoporoids were waning while the rhomboporoids and fenestrates were increasing. At this time the bifoliates and fistuliporoids reached their acme. In all, some 1000 bryozoan species, allotted to 175 genera, are members of the Devonian biota (Cuffey and McKinney, 1979).

Hemichordata

While a few(?) attached dendroid forms are known throughout the Devonian, true planktonic graptoloids are recorded only in the Lower Devonian. Some thirty Devonian species precede the demise of the group, monograptids for the most part and locally present in large numbers amongst the pelagic faunas world-wide.

Brachiopoda

The brachiopods reached into virtually every neritic habitat in Devonian time, achieving their acme during this period. Characteristic associations and communities have been recognised and the stratigraphic value of many of these is high (see figure 5.4). Many local zones based on lineages or on assemblages are distinguished and Boucot *et al.* (1968) acknowledge three world realms as for the rugosa. Among the Articulata the orthids were diminishing in number throughout the period but the strophomenids were still very numerous, the productid forms making their debut well before its end. The spirifers were at their peak in both numbers and diversity. Pentamerids were considerably reduced in number from the Silurian but both the rhynchonellids and terebratulids were on the increase. The Inarticulata are conservative and inconspicuous.

Mollusca: Gastropoda

Gastropod numbers were perhaps appreciably larger than in the Silurian. Benthonic forms were common and widespread and appear also to have become established in brackish waters. Fine-grained limestones with *Murchisonia*, *Loxonema* and others occur in many carbonate assemblages. *Bellerophon* and other pelagic forms (if acknowledged as gastropods) were widespread throughout the period (see figure 5.5).

Mollusca: Bivalvia

This group also seems to have spread from marine into brackish and fresh-water environments by Devonian time. They became widespread and diverse: some 35 families and more than 120 genera are known. Most remarkable was the explosive evolution of the Pterineidae and the Pterinopectinidae, groups which became important members of the nekton. Important bivalve communities in neritic environments in the Appalachian Devonian have been described (Bowen *et al.*, 1974; Thayer, 1974), but the bivalve faunas of carbonate facies have not been accorded the same attention.

Mollusca: Cephalopoda

Cephalopods from Devonian formations contribute one of the great evolutionary events in the history of the marine invertebrates. From the oldest Devonian rocks straight (or partly coiled or curved) orthoceratids and oncocerids are the sole cephalopod fossils found so far. Shortly afterwards the evolution of ammonoids from bactritoid straight-coned forms took place and during Middle and Late Devonian time the ammonoids underwent a burst of diversification. The true nautiloids, however,

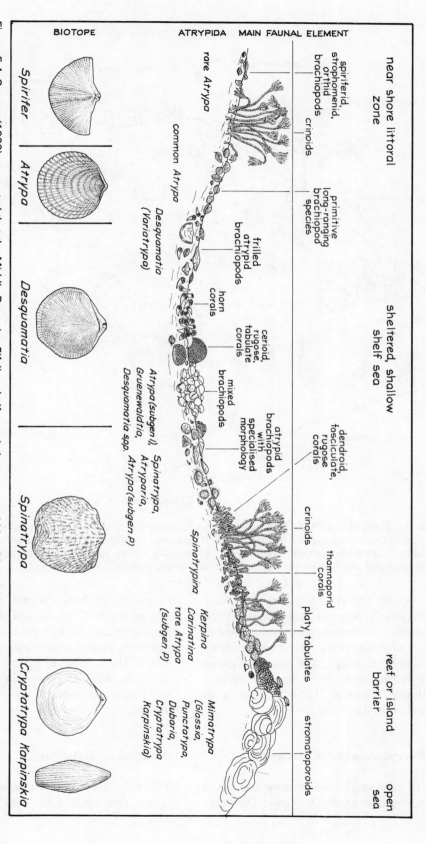

Figure 5.4. Copper (1966) suggested that the Middle Devonian Eifelian shelf revealed seven atrypid brachiopod biotopes or assemblages between the littoral zone and the open sea. Here a simplified version of that arrangement lists the common brachiopod genera found. The innermost biotope is essentially a 'Rhenish' magnafacies and the outermost is 'Hercynian' in the sense of Erben (1962).

BIOTOPE — ATRYPIDA MAIN FAUNAL ELEMENT

near shore littoral zone
- spiriferid, strophomenid, orthid brachiopods
- crinoids

rare Atrypa

sheltered, shallow shelf sea
- primitive long-ranging brachiopod species
- frilled atrypid brachiopods
- horn corals
- cerioid, rugose, tabulate corals
- mixed brachiopods
- atrypid brachiopods with specialised morphology
- dendroid, fasciculate, rugose corals
- crinoids

common Atrypa

Desquamatia (Variatrypa)

Atrypa (subgen I), Gruenewaldtia, Desquamatia spp.

Spinatrypa, Atryparia, Atrypa (subgen P)

reef or island barrier
- thamnoporid corals
- platy tabulates
- stromatoporoids

Spinatrypina

Kerpina (Glassia), Carinatina, rare Atrypa (subgen P)

open sea

Mimatrypa (Glassia), Punctatrypa, Dubaria, Cryptatrypa, Karpinskia

Spirifer

Atrypa

Desquamatia

Spinatrypa

Cryptatrypa Karpinskia

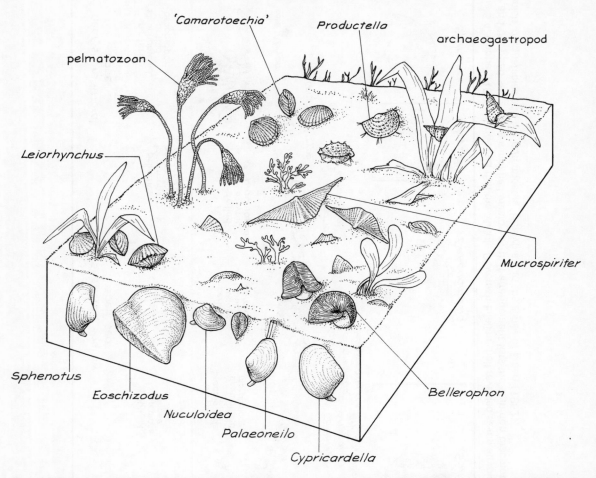

Figure 5.5. Amongst the four Late Devonian marine communities of the Sonyea Group in New York State is the *Bellerophon* Community with both infaunal and epifaunal members as shown (after Bowen *et al.*, 1974). The enclosing facies is largely delta platform and distributary mouth bars, channels and estuaries, the first of which also provided the habitat for the communities of greatest taxonomic diversity.

remain very poorly known. They appear to have been a conservative group and although about 250 Devonian genera are recorded, relationships within the group are obscure. While the ammonoids all appear to have been pelagic, the nautiloid group may have included forms that were inclined to a partly benthonic existence. Little is known of Devonian cephalopod ecology.

Mollusca: Dacryoconarida (Conocarida)

This group of very small shell-bearing animals is very well represented amongst pelagic Devonian faunas. X-ray studies suggest that it may be closely related to the Cephalopoda. It evolved rapidly in the Early and Middle parts of the period and is extremely abundant in pelagic facies throughout the world. The number of genera is not large; *Tentaculites*, *Styliolina* and *Nowakia* are commonly listed, but the group did not survive into the Carboniferous.

Arthropoda: Trilobites

Trilobites are found in most Devonian marine faunas (see figure 5.6). Some twenty families, in-

	STAGES	AGE (million years)	Duration (million years)	First appearance	Total present	Destined for extinction	Surving into the next stage
Upper		— — — — 345 — — —					
Upper	Famennian		8	11	23	15	8
Upper		353					
Upper	Frasnian		6	6	28	16	12
		— — — — 359 — — —					
Middle	Givetian		6	4	53	31	22
Middle		— — — — 365 — — —					
Middle	Eifelian		5	25	119	70	49
		— — — — 370 — — —					
Lower	Emsian		4	34	141	47	94
Lower		374					
Lower	Siegenian		16	51	123	16	107
Lower		390					
Lower	Gedinnian		5	?	88	16	72
		— — — — 395 — — —					

Figure 5.6. Trilobite genera occurring in the Devonian period. The numbers are approximate but give an indication of the fluctuations in trilobite diversity during some 50 million years (after Alberti, 1979).

cluding 107 genera, have been identified from these rocks, but the dominant group is the phacopids. A few forms were highly 'ornamented', spinose; they are bizarre and may have been pelagic rather than benthonic. Trilobite distribution is wide and faunas generally cosmopolitan; however, Ormiston (1972) has distinguished four trilobite provinces in North America that do not obviously conform to those of the corals or brachiopods. By the end of the period the superfamilies Calymenacea, Illaenacea, Phacopacea, Cheiruracea, Odontopleuracea and Lichacea had become extinct. Good world-correlation on the basis of trilobites is generally available and is improving.

Arthropoda: Ostracods

Four of the five orders of ostracods are present in Devonian rocks. Some ostracod families towards the end of the Silurian period were established in non-marine environments, the marine families being generally restricted to shallow-water environments. Devonian fresh-water ostracods are locally numerous; others include giant benthonic(?) species and open-water pelagic types which have in recent years been recognised as stratigraphically useful.

Towards the end of the period about 50 species of pelagic ostracods were common but only 16 of them survived into the Carboniferous.

Arthropoda: Merostomata etc.

Eurypterids, conchostracans and phyllocarids were plentiful during Devonian times, the first two categories also being found in non-marine facies. Other groups were appearing for the first time for the period was one of innovation and expansion for arthropods (see figure 5.7).

The Lower Devonian Rhynie Chert of Scotland (Westoll, 1977) and the Alken shale of West Germany give an insight into the crustacea, arachnids and collembola of that time. Eurypterids from Silurian rocks are well known but the Early Devonian examples are more diverse and the predaceous pterygotous forms were almost 2 m long. From the Hunsrück Slate pyritised pycnogonids and other arthropods have been studied using X-ray methods. At Alken a fossil scorpion, an arachnid and a myriapod are terrestrial arthropods, and some eurypterids may have occasionally ventured on to land; the invasion of the land by arthropods was well under way (Størmer, 1977).

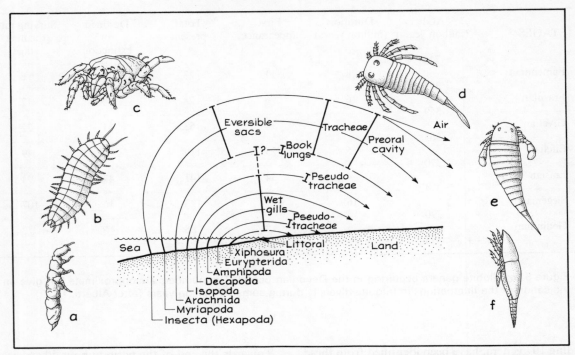

Figure 5.7. Late Silurian—Early Devonian times saw the invasion of the land by different arthropods, some of which are shown here: (a) the hexapod *Rhyinella*; (b) myriapod *Eoarthropleura devonica*; (c) arachnid *Protocarus*; (d) eurypterid *Drepanopterus*; (e) eurypterid *Parahughmilleria*; (f) xiphosurid. Størmer's (1977) suggestions of the respiratory adaptations needed for this are shown. Eurypterids are by far the most common of these fossils, but the other forms shown and several scorpions (not illustrated) occur in the Early Devonian.

Echinodermata

The echinodermata underwent a general reduction in range of anatomical organisation during this period. The Homalozoa became extinct; the Asterozoa seem to have left very few fossils, a puzzling fact in view of their record before and subsequently; the Echinozoa are rather scantily represented by five classes but the Crinozoa make the biggest showing by far. Three classes of blastoids are known and in some localities these pelmatozoa are abundant. The crinoids are especially abundant and diverse.

Conodonts

Great interest attaches to these phosphatic microfossils largely because of their problematic affinities, their almost universal distribution in marine for-

mations and their stratigraphic sensitivity (see page 78). They were at their acme during the period, several hundred species being known. Fahreus (1976a, 1976b) claimed evidence for two conodont provinces, Tasman–Cordilleran and Aurelian (southern Europe and Turkey) in the Lower Devonian which were reduced after Emsian times and eliminated by Frasnian time. A more detailed analysis of Devonian conodont endemism and dispersal has been made by Klapper and Johnson (1980). Conodont ecology has been treated by Fahreus (1976b), and Devonian conodont biofacies by Sandberg (1976) and Sandberg and Ziegler (1979).

Vermes and Trace fossils

Worms of many kinds were known from the pre-Devonian fossil record but perhaps the most commonly represented in the Devonian are the polychaetes in the form of scolecodonts (Crimes and Harper, 1970). Serpulid worms occur in many formations, usually as isolated individuals or small groups, but in Arizona a large biohermal mound of serpulid tubes is reported.

Trace fossils occur in many facies of the Devonian, showing that all the behavioural types were present in the marine realm (see figure 5.8). Undoubtedly a large number of the marine trace fossils represent worms of one kind or another. Several kinds are known from continental rocks, including the huge *Beaconites* from Antarctica and elsewhere and from the Canadian Arctic Islands (Narbonne *et al.*, 1979). A special study of the *Diplocraterion* trace fossil assemblage of the Late Devonian of North Devon has become something of a classic (Goldring, 1962).

Vertebrates

The earliest vertebrate remains are Late Cambrian and the earliest comparatively well-known agnathan is mid-Ordovician. In the Silurian the agnatha seem to have advanced very little but the antiarchi, arthrodira, acanthodians, osteichthyes and chondreichthyes all appeared. With the advent of habitable non-marine environments over a large part of the world in Devonian time all these groups but the last (see figure 5.9) appear to have made great strides (see Westoll, 1979).

Agnath and gnathostome groups were present throughout the period but the latter group rapidly assumed dominance in the Middle Devonian and by the end of the period virtually all forms but the major gnathostome divisions became extinct. In the Late Devonian the first tetrapods were in existence. The size attained by most of these vertebrates was exceeded by those in the seas where the arthrodires included forms (*Dinichthys* for example) that have never been surpassed in bulk.

The marine fauna may have been dominated by chondreichthyan groups (sharks, rays etc.) which are known from shales and limestones in North America, Asia, North Africa and Europe. A few

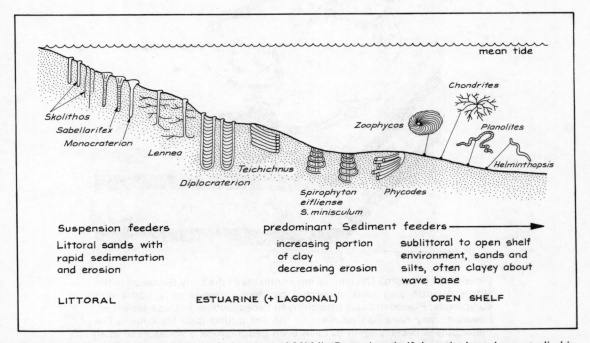

Figure 5.8. Trace fossil assemblages in Lower and Middle Devonian shelf deposits have been studied in several regions, notably the Rhenish Schiefergebirge and the Appalachians. This synopsis, from Germany, shows the commonest forms. Energy levels decrease with depth but trace fossil associations do not always reliably indicate the depth of overlying waters (adapted from Goldring and Langenstrassen, 1979).

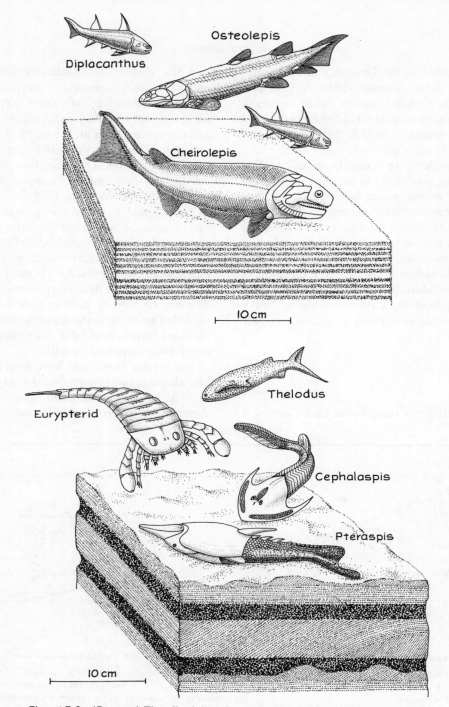

Figure 5.9. (Bottom) The alluvial environments of the Early Devonian in the North Atlantic area were inhabited by many species of agnatha and eurypterids. Placoderm and osteichthyan fishes are rare in these facies: all, however, may have had recourse to the sea during their life-cycles. The thelodonts especially seem to have been geographically widespread at all stages of their existence. (Top) A Middle Devonian lacustrine community in the Orcadian Basin may have included, apart from the forms shown above, antiarchi, arthrodires and lungfish. From time to time mortality was locally very high and perhaps because of desiccation, water bloom or other causes. The absence of traces of invertebrate animal life is puzzling (with apologies to W. S. McKerrow and R. Goldring—see Goldring (1978)).

agnatha are known from marine as well as continental formations.

Little is known about the mode of life of most of the Devonian vertebrates. The agnatha appear to have been relatively sluggish benthos for the most part, probably filter feeders or grazers (Halstead, 1973); the arthrodires and antiarchi included more active but also largely benthonic forms. The remainder were probably relatively active and predatory. Palaeoecological studies of Devonian vertebrates are very few. The limited use of vertebrates in Devonian continental biostratigraphy is referred to in chapter 7.

Correlation

Devonian biostratigraphy has been making great progress in the last decade or so, with an emphasis on micropalaeontological methods. This has not been entirely due to the needs of the petroleum industry although commercial support for academic studies has been important. The principal pelagic fossils pressed into service are the ammonoids, conodonts, dacryoconarids, graptolites, ostracods and palynomorphs.

Benthonic faunas of stratigraphic sensitivity include those of brachiopods, trilobites and to a lesser extent corals. The correlation of marine facies is approaching a relatively satisfactory state but the correlation of marine with non-marine formations is far less so. In this connection plants and spores appear to be the most useful groups, while the distribution of vertebrates in various continental facies and regimes provides a good but rather broad means of correlation. Problems of vertebrate identification and taxonomy, and stratigraphic and geographical distributions, are many.

Correlation by brachiopods

The most conspicuous and dominant forms in the Devonian benthos were brachiopods, the total number of described taxa being very large (see page 68) and increasing. The usefulness of these fossils, even though they are quite susceptible to facies control, was established in the early days of Devonian biostratigraphy. In Europe the brachiopod

facies have not only proved biostratigraphically useful but have served to identify faunal provinces. In more recent years phylogenetic studies have established the value of brachiopod lineages in Devonian stratigraphy. Nevertheless, time-scales based on brachiopods remain parachronologies rather than orthochronologies. While range zones of index fossils and assemblage zones have been recorded from many localities few well-defined boundaries can be fixed on the basis of these. The discussion of brachiopods in Devonian biostratigraphy by Johnson (1979) sets out the achievements and limitations of the work in this field so far.

In Europe the spiriferids have long been taken as index fossils as follows:

Acrospirifer	Siegenian–Lower Eifelian
Paraspirifer	Emsian–Eifelian (Couvinian)
Spinocyrtia	Eifelian–Givetian
Tenticospirifer	Frasnian
Cyrtospirifer	Frasnian–Famennian
Syringothyris	Upper Famennian–Carboniferous

Rhynchonellid brachiopods have also been advocated as good biostratigraphic index fossils in North America.

Different coeval brachiopod communities are known to exist side by side in many shelf areas and the transgressive–regressive nature of Devonian seas has led to the diachronous overlap of parallel brachiopod community 'zones'.

Allopatric speciation seems to have been the dominant mode of brachiopod evolution, but as Devonian seas transgressed the greater uniformity of marine environment in the continental margin reduced the rate and extent of local speciation. Thus brachiopod diversity was much reduced in the Late Devonian. The atrypids, orthids, gypidulids and stropheodontids were eliminated during what was the greatest extinction of marine invertebrates since the end of the Ordovician.

Correlation by graptolithina

Although the benthonic dendroid graptolites existed throughout the Devonian period and into the Carboniferous, the true graptolites have so far only been found within the lower series of the Devonian system. It was little more than 20 years ago that the first Devonian true graptolites were discovered. Since then they have been found in all continents except Antarctica and South America. As pelagic organisms they possessed fragile skeletons that have been preserved in the finer sediments of the Devonian orogenic belts. They have been more commonly found in some of the fine-grained clastic and carbonate rocks of the Hercynian facies than in the Rhenish in Europe (see Jaeger, 1979).

In all, some 30 Devonian graptoloid species and subspecies are known, divided among four genera. By far the commonest are species of *Monograptus* (see figure 5.10). They are relatively easily recognised as *Monograptus*, being small to medium sized with commonly between twenty and forty thecae to each rhabdosome. Most are robust and straight. By contrast, the rhabdosomes of *Lino graptus* and *Abiesgraptus* are multibranched and may be giants with several thousand thecae.

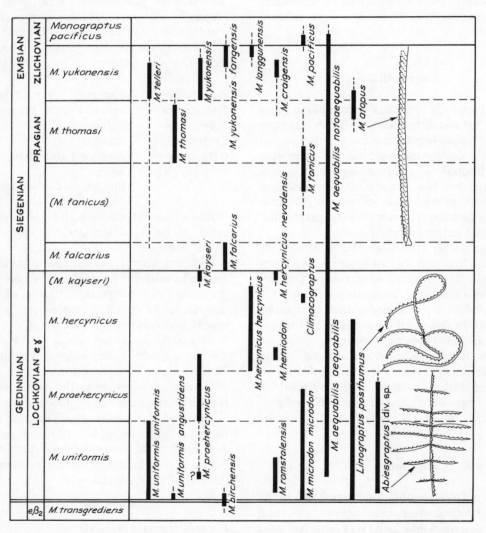

Figure 5.10. Ranges of Devonian graptoloids after Jaeger (1979), based on material from North America, continental Europe and North Africa, central and south-eastern Asia and Austraila. The zones of *Monograptus* listed on the left are based upon the ranges of individual species and only those of *M. uniformis* and *M. yukonensis* have more than five other monograptid species present.

Six graptolite zones have been recognised in all the continents where these fossils occur. The lower three zones (Lochkovian) are well established in many European sections and elsewhere. The upper three are less well substantiated. In several regions the *yukonensis* zone is clearly the latest and in some areas there are problematic unfossiliferous sequences below this zone. At the moment it seems that the correlation of graptolite ranges with those of other pelagic fossils is proceeding well, conodonts and ammonoids occurring locally with them. The evidence indicates that the last graptolites became extinct within the Zlichovian stage of the Bohemian (Hercynian) facies and within the Emsian of the Rhenish.

Correlation by dacryoconarida

Until relatively very recently these minute fossils were largely ignored for stratigraphic purposes. Now they are recognised as a group of major international importance. The pioneer studies were largely by Bouček (1964) in Czechoslovakia and his zonation has been found to be applicable in many other areas (Lütke, 1979)—see figure 5.11. Dacryoconarids are fossils of semi-microscopic size, simple chambered and annulated slender cones, found in both neritic and pelagic facies. They have proved very abundant in many fine-grained clastic successions. Phylogenetic lineages are being established for a number of stock but the zonation is

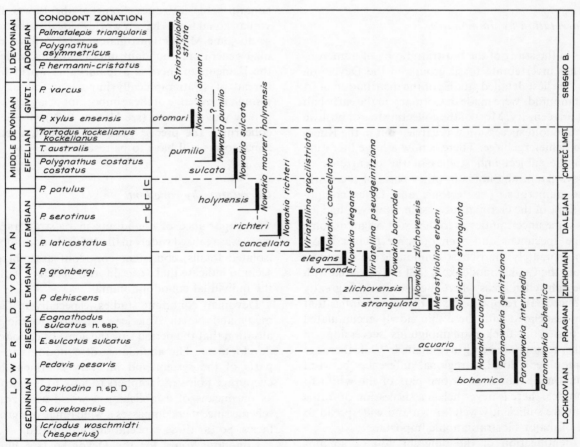

Figure 5.11. Dacryonarid zonation of the Devonian offset against the conodont stratigraphy as used by Lütke (1979). Only the fairly well-documented forms are shown here: incompletely known forms are not included, but they are in fact rather more numerous than those shown here. The lower boundaries of the Middle and Upper Devonian shown above are those that have been used by several conodont workers in the past (see Lütke, 1979; Klapper and Ziegler, 1979). Note that the full range of *N. zlichovensis* is uncertain.

readily determined on the basis of species or sub-species of the genus *Nowakia*. Twelve zones span the system from Late Gedinnian to mid-Upper Devonian. Fortunately the dacryoconarid faunas occur largely in facies that also yield conodonts so that a comparative chronology of the two may be attempted. The full ranges of all 41 or more dacryoconarid taxa known from the system are not fully established in terms of the conodont zones but nearly half the number is fairly well documented. Studies in Morocco and France, North America, Western Europe, Australia and China are in progress and seemingly testify to the world-wide spread of these fossils and their increasing importance to Devonian biostratigraphy.

Correlation by ammonoids

As in the study of the biostratigraphy of numerous other invertebrate fossil groups in the Devonian, the earliest detailed studies of the distribution of its ammonoids were made in Germany in the early half of this century. Most of the collecting to establish an ammonoid succession was carried out in the Rhenish Schiefergebirge. There is now a large literature on the subject and studies of the occurrence of Devonian ammonoids in other parts of the world are numerous. The tectonic and facies complications of the German succession, however, had led to some uncertainties as to the exact origin of some type specimens and even of some faunas. Most unfortunately the richest ammonoid faunas come from the Cephalopodenkalk that is associated with the Schwellen areas where the succession is greatly reduced. To resolve some of the uncertainties that are associated with these thin slowly accumulated deposits the thicker unambiguous successions in other regions are being investigated. While there are undoubtedly local or regional differences between ammonoid faunas from one part of the world to another, there is nevertheless a succession of forms that are sufficiently well known and widespread to be of major biostratigraphic importance.

Identification of the different 130 or so ammonoid genera within the Devonian has almost entirely been made on the basis of the shape of the suture in the adult, and most ammonoid zones have been based on genera rather than species (see figure

5.12). In recent years additional criteria such as shell form and ornamentation have also been utilised.

The earliest ammonoids known in the Lower Devonian are loosely coiled forms, suggestive of an ancestry from a curved bactritid, and from these there appears to have been a vigorous evolution into a number of stocks that typify the late Lower Devonian and the Middle Devonian. Most of these die out, to be replaced in Frasnian time by other groups that are in turn largely replaced in the Famennian.

The German practice has been to divide the Devonian into 8 broad biostratigraphic units, *Stufen* (see figure 5.12), named after key genera and containing in all some 20 discrete zones. In North America the rich faunal succession is found to contain some 30 zones. Some of the Lower Devonian zonal forms have a world-wide distribution, as do some Middle Devonian genera. A few Frasnian genera are abundant internationally while in the Famennian there is a remarkable world-wide spread of stratigraphically important forms. Although knowledge of these important pelagic fossils is now very extensive (see House, 1979), problems concerning the precise range of some distinctive zonal forms still have to be resolved.

Correlation by conodonts

The importance of conodonts in biostratigraphy work was realised widely in the 1950s. In addition to isolated fossils, consistent conodont assemblages seem to indicate that several forms occurred within the individual conodont animal.

Devonian conodont studies, especially in Germany and North America, led early to the realisation that the period saw the acme of conodont development. The abundance of conodonts in all parts of the system and in every continent has encouraged biostratigraphers to look to this group as the means of establishing a world-wide zonal scheme since it encompasses a wide range of marine facies. So far these expectations are upheld.

Conodont zones for the Devonian (see figure 5.13) rely particularly on the phyletic lineages of a relatively small number of genera. They have been proposed for Devonian sections in many parts of the world and the correlation of these sections has

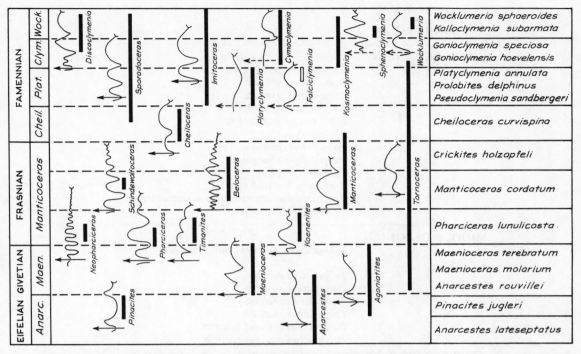

Figure 5.12. The ranges and diagnostic sutures of the ammonoid genera that are useful in zonation of the marine Devonian-named zones are on the right, the 'Stufen' on the left. See figure 4.4 for a more complete listing of the European and North American zones, stages and faunas (after House, 1975b).

been a major feature of Devonian conodont studies in recent years (see Klapper and Ziegler, 1979). From this work it has become apparent that lower Lower Devonian conodont facies in western North America differ substantially from those of Western Europe. The icriodids, *Ozarkodina* and related genera are the basis for this differentiation, while upper Lower Devonian and Middle Devonian zones, however, tend to be more universally recognisable, extending to Australia, though some icriodids are locally endemic. In the upper Middle Devonian the endemic elements further diminish and in the Upper Devonian the zonation seems to be consistent world-wide. These higher zones of the Devonian are excellent examples of the use of phyletic lineages in biostratigraphy. The large platform-type conodonts involved are *Polygnathus–Schmidtognathus*, *Polygnathus–Palmatolepis*, *Polygnathus–Ancyrodella–Ancyrodella–Ancyrognathus*, *Bispathodus* and *Siphonodella*. Subzones are distinguished at a number of levels,

the *Polygnathus varcus* zone being subdivided into three, for example.

Despite our continuing ignorance of the nature of the conodont animal, it has been possible to discern in the distribution of coeval conodont faunas the 'preference' that some taxa exercise for particular lithofacies and their absence from others in the Middle and Upper Devonian. For example, Sandberg (1976) has in the Rocky Mountains found five laterally equivalent but different late Upper Devonian faunas. Distinguished as conodont biofacies, they are characterised by different percentages in the association and morphology of species:

1. Palmatolepid–Bispathodid biofacies
   continental slope and rise
2. Palmatolepid–Polygnathid biofacies
   continental shelf
3. Polygnathid–Icriodid biofacies
   moderately shallow water on outer shelf

CARBONIFEROUS

UPPER DEVONIAN
- *Siphonodella praesulcata*
- *Bispathodus costatus*
- *Polygnathus styriacus*
- *Scaphignathus velifer*
- *Palmatolepis marginifera*
- *P. rhomboidea*
- *Palmatolepis crepida*
- *Palmatolepis triangularis*
- *Palmatolepis gigas*
- *Ancyrognathus triangularis*
- *Ancyrognathus asymmetricus asymmetricus*
- *Palmatolepis disparilis*

MIDDLE DEVONIAN
- *Schmidtognathus hermanni-cristatus*

- *Polygnathus varcus*
- *Polygnathus ensensis*
- *Tortodus kockelianus*
- *Polygnathus australis*
- *Polygnathus costatus costatus*
- *Polygnathus costatus partitus*

LOWER DEVONIAN
- *Polygnathus costatus patulus*
- *Polygnathus serotinus*
- *Polygnathus inversus*
- *Polygnathus gronbergi*
- *Polygnathus dehiscens*

- *Eognathus sulcatus*

- *Pedavis pesavis (Icriodus)*

- *(Ozarkodina n.s.p. D.)*
- *Ozarkodina eurekaensis (I. postwoschmidti)*
- *Icriodus woschmidti hesperius (and I.w. woschmidti)*

SILURIAN

Figure 5.13. The stratigraphic ranges of a large number of Devonian conodont species are known in many parts of the world. Differences between the successions in the lower part of the Lower Devonian in western North America and in Western Europe are substantial but more cosmopolitan species are thereafter conspicuous. The figure gives those zones that are the most widely and easily recognised and is largely based upon the evolution of polygnathid platform species. Two or three subzones per zone are also increasingly recognised but they are not listed above (after Klapper and Ziegler, 1979).

4. Polygnathid–Pelekysgnathid biofacies
   shallow water, inner shelf
5. Clydagnathid biofacies
   very shallow variably saline waters over off-shore banks and in lagoons

While the outer biofacies contain 'normal' bi-laterally symmetrical species, asymmetrical and bizarre forms are dominant in the innermost.

Correlation by spores

Biostratigraphically sensitive groups of fossils are, as we have seen, numerous in the marine Devonian rocks. The continental facies, however, have few fossils in common with the marine but locally spores are among them. Not all facies of the continental Devonian contain spores, and by no means all spore taxa are widespread. The parent plants probably grew at the margins of bodies of fresh water, and the spores were adapted for dispersal by wind and water. Spores are known from the sporangia of about 60 species of Devonian plants and compare in morphology to about 17 genera of dispersed spores. The spores themselves are minute spherical bodies that are composed of the hardy material sporonin; they were produced in immense numbers by each reproductive body of the plants. They could survive under rigorous conditions and, where incorporated as grains or sediment, they commonly survived corrosion and carbonisation in large numbers.

Identification is based upon the size, shape, structure and ornamentation of the individual grain or palynomorph. In recent years these fossils have been found locally in sufficient numbers for zonal schemes to be proposed, and much information on spore assemblages, ranges and zones has been forthcoming, especially from North American, Western Europe and the U.S.S.R. (Richardson, 1974; McGregor, 1979). Their value is potentially high in that they appear to afford a means of correlating the continental with the marine facies from all parts of the Devonian system and possibly for all parts of the world.

Most of the palynological literature to date has referred to the Rhenish rather than the Bohemian stages of the Devonian. This is largely because spores are common in the Rhenish sections but well-preserved spores are common only in the Zlichov–Daleje–Trebotov beds of Bohemia. While we cannot expect spore zones to coincide with those of other organisms, the prime requirement once such zones have been recognised is to calibrate them with the zones of co-existent marine faunas. McGregor (1979) has identified those spore-based marine reference sections that seem best to comply with this in one part of the world or another. A tentative zonal scheme for the entire Devonian system of the northern hemisphere was offered by Richardson (1974) but as yet no such scheme of international spore zonation has been internationally accepted. McGregor (1979, fig. 8) has nevertheless compiled a chart that shows the ranges of 49 species of spores that have wide geographical distribution, with short stratigraphic spans, and that can be found in both marine and non-marine facies and recognised relatively easily even when poorly preserved. A further most critical point is that the ranges of these spores are known relative to associated marine faunas.

Concerning global patterns of spore distribution McGregor makes the following comments.

1. Although far more is known about the spores of the Old Red Sandstone Continent (North Atlantis) than elsewhere, at most times during the Devonian there were near-cosmopolitan species. Examples from each Devonian series fall within the approximate latitudes 30°N to 60°S.
2. Some spore assemblages appear in areas that are thought to have been even closer to the South Pole and this may indicate that at times within each Devonian epoch the climate of southern Africa was not rigorously polar (see figure 5.14).
3. Most of the more cosmopolitan species of spores have relatively long stratigraphical ranges and their parent plants may be assumed to have had a wide environmental tolerance. Thus from when the world-wide proliferation of land plants began early in the period there were widespread and long-lasting elements in the flora.
4. A few globally distributed species have the relatively short stratigraphic range of a stage or less.
5. Some species, however, have a very restricted geographical distribution, especially within the North Atlantic area.
6. Possibly distinct palaeofloral regions—North

	NORTHERN HEMISPHERE Richardson (1974)	EUROPEAN U.S.S.R. Chibrikova and Naumova (1974)
TOURNAISIAN		
	V. nitidus	X (*H. lepidophytus*)
	V. pusillites-S. lepidophytus	
UPPER DEVONIAN / FAMENNIAN		IX (*C. varicornata*)
	L. cristifer	VIII (*Archaeotriletes*)
FRASNIAN	*optivus-bullatus*	VII (*Archaeoperisaccus*)
		VI
	?	V (*Archaeozonotriletes*)
MIDDLE DEVONIAN / GIVETIAN	*triangulatus*	IV (spinose *Archaeozonotriletes*)
EIFELIAN	*D. devonicus*	
	R. langii-A. acanthomamm-illatus	III (*Retusotriletes-Hymenozonotriletes*)
EMSIAN	*C. biornatus-C. proteus*	II
	E. annulatus	
LOWER DEVONIAN / SIEGENIAN	*D. cf. gibberosus*	I
	?	
GEDINNIAN	*E. micrornatus-S. newportensis*	

Figure 5.14. Attempts at zoning the non-marine (O.R.S.) facies by spores involve much labour and painstaking work which, so far, have not been completely successful. Two of the many suggested schemes are shown. Universal recognition and acceptance of these zones has yet to be achieved (see McGregor, 1979).

Atlantis, Central Asia and medial South America — may be indicated by the differences in coeval assemblages but the hypothesis is as yet very tentative in view of the very few species that have so far been described from Central Asia and South America.

Biogeographic Provinces

The concept of biogeographic provinces confined by physical or climatic barriers that prevent the dispersal of organisms is well established for the present biosphere. It may be inferred from what we see today that similar barriers and provinces existed over the lands and seas of the past. The geological record confirms that animals and plants of the past had similar ecological relationships to those of their descendants and were distributed about the world in similar ways. The literature on the subject is now large (see for example Hallam (1974) for an excellent commentary and list of references and Boucot and Gray (1980) for further discussion and references).

Organic evolution has been constantly influenced by geological and hence environmental changes and, with the passage of time, barriers to the dispersal of organisms have come and gone, moved their positions and generally controlled the nature of biogeographical distributions. The biogeographic provinces of today have been modified from those of Pleistocene and earlier times by geological and organic evolution. In the same vein Devonian biogeographic provinces were developed from those of the Silurian period and in their turn were the basis from which those of the Carboniferous evolved.

Within each such province the general ecosystems and biological associations can be found throughout. From the fossil record we can establish the distribution patterns of organisms of the past; it is, however, much more difficult to recognise more than the vague outlines of the ecology within these areas. In recent years palaeoecology has made considerable progress and the palaeogeography of past times has been deduced in enough outline to allow reasonable reconstructions of climatic belts, oceanic circulation and other features that exert a controlling influence upon the distribution of life.

During 40 to 50 million years of the Devonian period there was a general northward shift on the continents and a shuffling of the continental blocks on the northern side of the Gondwanaland supercontinent. Orogenies changed the topography of wide areas, and volcanic activity occurred in many quarters. Coupled with these events was the changing level of the sea, which culminated in the great Late Devonian transgression. Such events had powerful effects upon the distribution of life and upon the biogeographic provinces of the time.

The first appearance of a new species is a unique biological event and the spread of the species throughout its biogeographic province is generally assumed to be so rapid compared to the pace of geological changes that the first record of the new species in sections throughout the province is thought to be everywhere synchronous. The extinction of the species, however, may be markedly earlier in some areas than in others. Extinctions are unreliable events by which to mark moments in geological time. The list of species or other taxa that survived locally well beyond the demise of those forms of life generally is a long one. Migrations and extinctions may be said to constitute only secondary time-significant events.

Palmer (1965) has suggested that migrations of trilobite faunas in Early Cambrian time can be seen to have taken place slowly over some 500 km of the shelf seas of the Cordilleran region of North America. Not all palaeontologists accept this and believe that is is not sufficiently clear which changes in fauna are due to the normal course of evolution and which are due to the incursion of migrants that were previously present elsewhere.

Very few statistical works have been undertaken in the field of palaeobiogeography: those to have appeared concern the recognition of the changing proportions of endemic and pandemic taxa in particular areas throughout the course of time. A 'provinciality index' that could be reliably obtained from the numbers of different taxa present in (or absent from) regions of perhaps different size is desirable. Simpson's (1960) well-known coefficient of similarity between two faunas is $(100\,CN_2)/N_1$ per cent, where N_1 represents the smaller number of fauna, N_2 the larger and C the number of taxa in common. Johnson (1971) in a study of the two Devonian brachiopod provinces of North America suggested a provinciality index $PI = C/2E_1$; here C is the number of taxa common to both regions and E_1 the smaller of the (two) numbers of provincial genera being compared. This gives provinciality for values of PI less than 1 and cosmopolitanism for values more than 1. It is a rapid visual evaluation of the position (see figure 5.15), as indeed is Simpson's method. Savage *et al.* (1979), in a quantitative analysis of Lower Devonian brachiopod distribution, proposed some useful refinements.

Faunal provinces as indicated by fossil evidence may be expected to include a range of faunal (or bio) facies, related to depth of water, nutrient and oxygen supply, turbidity, energy levels etc. One faunal facies may be sharply distinguished from those adjacent to it or it may emerge by slow change. Facies fossils can normally be recognised and distinguished from those that vary with the passage of time rather than change of habitat. Only when correlation between different facies has been achieved to some degree can the palaeontologists proceed to survey the distribution of equivalent faunas on the global scale. The study of Devonian faunal distributions is already influenced by ideas of plate tectonics and continental drift, and hence is still tentative. The most significant studies so far

have been those concerned with rugosa (Oliver, 1973, *et seq.*), brachiopods (Boucot *et al.*, 1968, 1969, *et seq.*), and vertebrates *(Young, 1981)*.

Devonian Faunal Provinces

Biogeographers of Recent, Cainozoic or Mesozoic life have begun to look afresh at ideas on species distribution and diversification. The theory of vicariance biogeography has been presented as more in keeping with the vast quantity of data now at hand than as ideas about species dispersal from centres of radiation or of distribution by wind, water or locomotion. An emphasis appears now to swing towards the idea of plate movements, eustasy and climatic change being responsible for great biogeographical changes since the Early Mesozoic. This 'panbiogeography' lays emphasis upon the splitting of once virtually universal and uniform realms of terrestrial diversification following isolation (see Nelson and Rosen, 1981).

Discussion of Palaeozoic biogeography remains limited and so far not much influenced by this, being for the most part confined to description of distributions in the shallow marine realm and of the

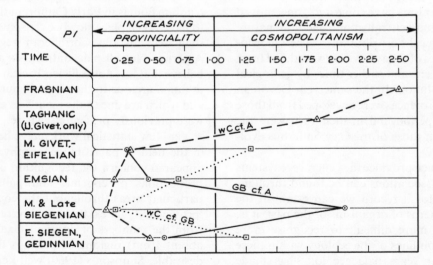

Figure 5.15. Provinciality Index (*PI*) of Devonian brachiopods plotted against time. The dashed line links *PI* values from a comparison of western Canadian and Appalachian genera; the solid line links those from a comparison of Great Basin (western U.S.A.) and Appalachian genera; the dotted line links those from a comparison of western Canadian and Great Basin genera (after Johnson, 1971; Johnson and Boucot, 1973).

Middle to Late Palaeozoic plants and vertebrates. Variance biogeography has yet to make its impact upon the topic. Marine biogeographic provinces in the Devonian are well substantiated.

The three European facies groups (or magnafacies), Rhenish, Bohemian or Hercynian and Old Red Sandstone, still present problems of correlation. The marine facies faunas also appear to differ from those in other parts of the world sufficiently to indicate the existence of faunal provinces early in the period. Endemism as typified by the brachiopods, corals and some other fossils of that time gave way to the rise of cosmopolitan faunas (Boucot *et al.*, 1968; Johnson and Boucot, 1973; Oliver, 1973, 1976).

The three major faunal provinces are (1) Old World Province, (2) Appalachian Province and (3) Malvinokaffric Province. The first two of these were already in existence at the beginning of the Devonian but they became more discrete; the endemism that defined them became more pronounced throughout Early and Middle Devonian times. The Malvinokaffric Province came into existence in Emsian time probably by migration of fauna from the Appalachian Province in northern South America (see figure 5.16). Thereafter Devonian provinciality diminished.

The *Old World Province* includes Europe and North Africa, parts of the Near East and parts of Nova Scotia as a *Rhenish–Bohemian subprovince*. A *Uralian subprovince* includes the Urals and the Asiatic region of the U.S.S.R., except Kazakhstan and parts of the Transbaikal region. The *Tasman subprovince* includes eastern Australia and the *Cordilleran subprovince* covers western North America.

The *Appalachian Province* covers much of eastern North America from the mouth of the St Lawrence to Texas and Colombia and Venezuela with connections to south Central Asia and perhaps eastern North Africa and Spain.

One of the palaeogeographic features that is strongly suggested by brachiopod distribution in

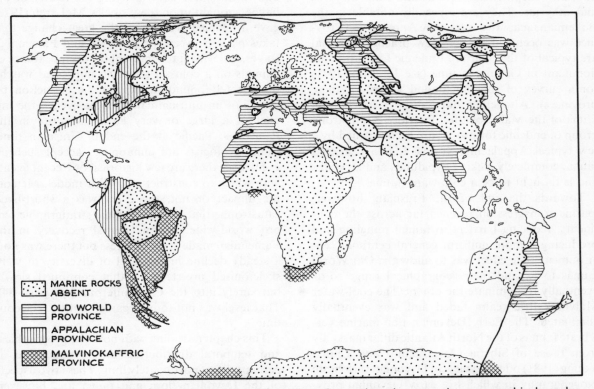

MARINE ROCKS ABSENT

OLD WORLD PROVINCE

APPALACHIAN PROVINCE

MALVINOKAFFRIC PROVINCE

Figure 5.16. Biogeography of Emsian stage brachiopods as plotted on a palaeogeographic base by Johnson and Boucot, 1973. Although the continents have not been 'restored' to their Devonian positions, the relative sizes and extents of the three provinces are clear.

North America is the Transcontinental Arch that separates the Old World Cordilleran subprovince from the Appalachian Province, which reaches from Ontario to Nevada. Appalachia, a product of the closure of Iapetus, separated the Appalachian Province from the Old World Province. Oliver (1973) showed that this barrier increased in effectiveness between Silurian and Emsian time but diminished in the Givetian; at which later time rugosan genera from the Old World Province in the Dakotas and Saskatchewan crossed the Transcontinental Arch into the (Old World) Michigan Basin. In Late Devonian time the arch was broken by several seaways which allowed the two faunas to mix.

The *Malvinokaffric Province*, named after the Falkland Islands (or the Malvinas), occupies South America south from Peru and the mouth of the Amazon, South Africa and Antarctica.

The brachiopod fauna of the Old World Province includes groups that are persistent from the Silurian—eospiriferids, orthids and some atrypids and athyrids. They are especially notable in the Bohemian facies. In contrast the Appalachian Province was occupied by very few brachiopods that are typical of the Silurian. Endemic forms became dominant in Gedinnian time (see Johnson (1971) for a survey of the brachiopod genera of these provinces). Altogether different, the brachiopod fauna of the Malvinokaffric Province has a strong group of endemic forms that are accompanied by a few typical Appalachian forms. It is a restricted fauna, completely lacking gypidulids and atrypids, and is thought to be a cold-water fauna.

Towards the end of the Frasnian, however, marine transgression reached far across the continents, reducing barriers to faunal mingling and producing a more uniform general benthonic environment. The effect was to allow the Old World faunas to expand their geographical range so as eventually to dominate the others. The cool-water Malvinokaffric realm faded and was eventually eliminated. The Early Devonian non-marine vertebrate faunas of the North Atlantic differ markedly from those of Siberia (Angaraland) and China. Young (1981) offered a view of Devonian vertebrate biogeography in which five early Devonian provinces occur: a cephalaspid province (Euramerica), an amphiaspid province (Siberia) with an adjacent tanuaspid province (Tuva), a galeaspid–yunnanolepid province (South China) and a wuttagoonaspid–phyllolepid province (East Gondwana). In each of these relatively isolated areas early vertebrates developed along different lines (see figure 5.17). At the end of Silurian time the faunas of Laurentia and Baltica were mingled and by late Middle Devonian time there was dispersal between Gondwana and Euramerica. Migration may originally have been hindered by deep waters, changes of temperature, salinity, land etc., but as time passed such barriers diminished in effectiveness. The bony fishes seem to have achieved an almost universal distribution by mid-Devonian time; endemic forms none the less arose in each of the major continental provinces. So far no verified reports exist of vertebrates in the Malvinokaffric waters of South America but they are known from the Middle Devonian of South Africa.

The exact causes of the extinctions amongst the Late Devonian marine invertebrates remain unknown but there is no denying the reality of the havoc wrought upon these stocks. McLaren (1970) gave a detailed account of the great change that takes place between the Frasnian and Famennian stages, believing it to represent the disappearance of animals on a colossal scale. So impressed was he that he felt bound to explain the extinction by means of an unusual catastrophic event—the impact of a large or very large meteorite in the Palaeozoic Pacific at the end of Frasnian time. Many geologists are unhappy to accept such an event since there are few historical or Recent events with which to construct a possible model. Meteoritic impact or not, the event was a short-lived catastrophe that has marked the stratigraphic record world-wide. Marine faunal recovery in the Famennian made good progress but there was now a steady decline in the level of diversity of well-skeletonised invertebrates that continued slowly but surely into the Carboniferous (see page 64). That level was not to be regained until Cretaceous time.

This chapter, dealing with both the geographical and temporal distribution of Devonian life, is perhaps disproportionately long. These two aspects of the Devonian flora and fauna are, however, inescapably linked—an important point in any consideration of the Devonian world. The distri-

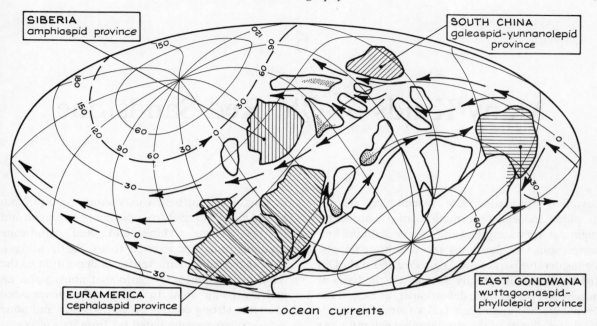

Figure 5.17. Major vertebrate faunal provinces of the Early Devonian lay within the tropical latitudes. The suggested oceanic circulation within this belt perhaps had by mid-Devonian time enhanced the development and spread of a cosmopolitan vertebrate fauna (see also figure 10.4). Within several of the seaways, moreover, mobile uplifts and island arcs temporarily helped to link these 'continental' provinces one to another. Stippled areas represent regions where non-marine vertebrates are also found between the major continental provinces.

bution of different forms of marine life was perhaps more intricate than previously, the environments being more varied and expanding in the non-marine realm. Biostratigraphically useful taxa similarly appear to be more numerous than for earlier systems.

6

Many Kinds of Old Red Sandstone

Devonian non-marine facies assemblages are not the oldest Palaeozoic formations of a continental origin but are widespread and they contain the first conspicuous vascular floras and vertebrate faunas. A considerable part of the Anglo–Welsh Lower Old Red Sandstone is actually Silurian on the basis of the new biostratigraphic definition of the base of the Devonian. Although such facies form only a relatively small part of the stratigraphic column most have been preserved in rapidly subsiding basins that are adjacent to areas of orogeny or uplift. Many are post-orogenic molasse resting upon deformed basement. The Old Red Sandstone of the British Isles represents for the most part alluvial and lacustrine associations; there are many others. The study of these continental sediments contributes an invaluable part of the Devonian record throughout the world: only South America seems to be without an extensive Devonian continental record. This distinctive magnafacies is, however, also found in Late Silurian strata (Britain, Canada) and may extend into the Carboniferous in many regions (Europe, Greenland, U.S.A.). The last decade has seen a surge of interest in these rocks, and many local sedimentological models have been proposed (see Van Houten, 1977). Nevertheless few of them are complete in all respects and many are disputed. Their exact ages are in many cases far from certain.

To begin with, continental sedimentary basins may be broadly spoken of as *internal*, or having little or no access to the ocean, or as *external*, where they extend into the marine (oceanic) realm. While this is roughly a physiographic distinction it may also be a tectonic one. Internal basins may be expected to show a (greater) preponderance of coarse clastic lithologies and sedimentary body shapes that are indicative of terrestrial debris cones

or fans. External basins may show the unbroken lateral chain of facies from piedmont to deltaic and inshore marine. Many Old Red Sandstone accumulations are referred to as clastic wedges, their shape being characteristic of deposition on the flank of a sedimentary basin that abuts upon an actively rising area. In all of them provenance exerted a strong control on composition and none seems to have accumulated far from its sources. It remains something of a puzzle that red alluvial rocks are not more evident in the Lower Palaeozoic systems.

Selly (1970) noted the occurrence of rocks of alluvial origin as mainly:

1. Prisms (or wedges) of sediment, thousands of metres thick, that are deposited in basins adjacent to mountain chains.
2. Sequences or masses, thousands of metres thick, that are deposited in fault-bounded troughs within continental shields.
3. Laterally extensive sheets of coarse, braided alluvium, generally only a few hundred metres thick that rest on continental shelf.
4. Thin sheets of alluvium that are beneath transgressive marine deposits.

These divisions are of course not mutually exclusive; one may grade laterally into another.

The first type (as in an *external* basin) is exemplified by the Catskill facies assemblage in North America and by the Old Red Sandstone of South Wales.

The Old Red Sandstone of western Norway, the north of Ireland and the Midland Valley of Scotland, typical of *internal* basins, occurs in fault-bounded valleys *within* the Caledonide mountain chain and the Orcadian Old Red Sandstone may be an example of the second type, although its original limits are still uncertain.

The third type of deposit is represented by the extensive Late Devonian red beds in Eastern Europe, eastern Australia and Antarctica.

The remaining type, thin sheets beneath marine transgressions, occurs widely beneath the marine Devonian of Western Europe and western North America, and in many parts of the Devonian of central and eastern Asia.

Red pigment, common in ancient continental deposits of almost every kind, is also present in many transitional and neritic facies, although is not common in recent alluvia. Argument about its origin—whether detrital or diagenetic—has been waged for many years (see Turner, P., 1980). Recent subaerial deposits redden with age, and there seems to be no doubt about the efficacy of the diagenetic reddening processes in tropical and semi-tropical environments. The breakdown of ferromagnesian minerals under temperate continental conditions, too, seems to produce red pigments comparatively rapidly.

A practical classification of continental red beds devised by Turner, P. (1980) seeks to take into account all their significant characteristics (figure 6.1). The initial depositional environment exerts an important influence on subsequent diagenesis. Thus alluvial, desert and delta plain red beds are readily distinguished on the basis of the characteristic facies

	Depositional characteristics	Redness characteristics	Diagenetic characteristics
ALLUVIAL RED BEDS			
Alluvial braided rivers (Proximal facies)	Pebbly alluvium. Mostly multistorey channel deposits, interbedded debris-flow and stream flood deposits	Uniformly red. Finer-grained horizons more intensely red	Intrastratal solution of silicate grains. Increase in mineralogical maturity. Decrease in textural maturity. Authigenic quartz, feldspar, calcite, clays
Alluvial plains with high or low-sinuosity streams (Distal facies)	Sandy and muddy alluvium in FU cycles. Sand units are channel and bank deposits. Mudstones are flood-plain deposits. Pedogenic modification of flood-plain deposits common	Variegated. Sandy units red or drab. Mudstone units red or variegated	As above. Also inversion of ferric hydroxides—haematite in muddy flood-plain deposits. Calcrete a common feature of flood-plain deposits
DESERT RED BEDS			
Alluvial, aeolian sabkhas, desert lakes	Cross-stratified well-sorted sands with steep foreset inclinations. Interbedded with inland sabkha and desert lake (gypsum and anhydrite) and poorly sorted wadi deposits	Uniformly red	Intrastratal solution of feldspars and ferromagnesian minerals. Authigenic quartz and feldspar. Gypsum–anhydrite cements
DELTA PLAIN RED BEDS			
Delta plains of river-dominated deltas	CU and FU cycles. Siltstones frequently associated with seatearths and other pedogenic modifications. Mainly flood-plain, well-drained swamp and lacustrine delta deposits	Variegated. Red beds usually confined to mudstones. A mottled appearance is common	Early diagenetic pyrite. Formation of kaolinite by humid acid leaching. Siderite. Pedogenic nodules

Figure 6.1. A classification of red beds into major associations based on depositional environment and colour characteristics (after Turner, P., 1980). CU=coarsening upward, FU=fining upward.

details. Red bed deposition seems on the basis of palaeomagnetic data to have been most common between the palaeolatitudes of 20°–40°.

As outlined above (page 28), classifications of depositional environments are many. The geologist may consider them in terms of physical, chemical and biological criteria or in terms of the deposits that accumulate within them. Each environment is characterised by a definite set of conditions that influence the sediments deposited. A simple classification of non-marine environments is

		DEVONIAN EXAMPLES
Fluvial	Piedmont Intermediate Flood plain	Scotland, Greenland and Spitsbergen
Desert	Enclosed basins Wide low-lying deserts	Scotland and Southern Ireland

Swamp	Basins with rigorous
	    humic
	    sedimentation
		Canadian Arctic
		Islands

Lakes	Shallow
	    lakes
	    and
	    playas
	Deep-water lakes
		Canada, the North
		Sea, and Altai-
		Sayan, U.S.S.R.

Glacial	Ice-related
	Outwash plains
	    and fans
	Aeolian sheets and
	    dunes
	Lakes
		Parnaiba Basin(?)
		and north-east Brazil
		Andean Basins
		and Bolivia

Most of these environments are more diverse and variable than are those of the marine realm. The boundaries between terrestrial environments tend to be sharper than those of marine environments and many physical conditions vary not only seasonally but even daily. Perhaps the most important physical factor to influence the nature of the non-marine environments is the relative abundance or absence of water since it affects the biosphere, weathering and the transport of sediment.

Conybeare's (1979) table of continental environments and their associated depositional models includes the common component physiographic–sedimentological features (see figure 6.2). The latter are all-important in determining the shapes and

internal structures of the major bodies of sediment that constitute the continental facies. Most of them are well known in Recent analogues.

Recently A. D. Miall of the University of Toronto made a comprehensive analysis of alluvial basins, noting that they tend to be elongate parallel to structural grain, or to occupy cross-cutting graben systems (Miall, 1981). Models based on the drainage patterns that develop in such basins (see figure 6.3), and on the type of depositional environment, reflect tectonic controls and occur in twelve major tectonic settings for which Miall found convincing modern analogues.

Studies of modern continental environments have been numerous in recent years so that the origins of distinctive lithological features, sedimentary structures and the geometry of sedimentary bodies are now better understood. As Van Houten (1977) has reminded us, three facets of interest have come to dominate the recent work on detrital sediments and sedimentary rocks: hydrodynamics of sedimentation, flow regimes and associated bedforms; deltaic and tidal deposits; and statistical analysis of ancient deposits. Each contributes to a better understanding of the sedimentary rocks and of the palaeogeographies within which they originated.

General Facies Characteristics

Within the classic Old Red Sandstone of the North Atlantic area and in the continental Devonian elsewhere the characteristic facies include aeolian, proximal median and distal alluvial and debris-flow sediments and they extend to marine-influenced types such as coastal mudflats, coastal barriers and inshore facies. Terrigenous clastics predominate; some are extremely coarse grade although the bulk are within the grades from coarse gravel to silt. Carbonates, present as pedogenic caliches, pedocals or calcretes, are now known from such widespread localities as Spitsbergen (Friend and Moody-Stuart, 1970), Europe, China (Boucot *et al.*, 1982), Antarctica (McPherson, 1979) and Australia. Associated with lacustrine environments are stromatalitic and other limestones. Where a marine influence is present the common red and green colours give way to greys and yellows. Sedimentary structures

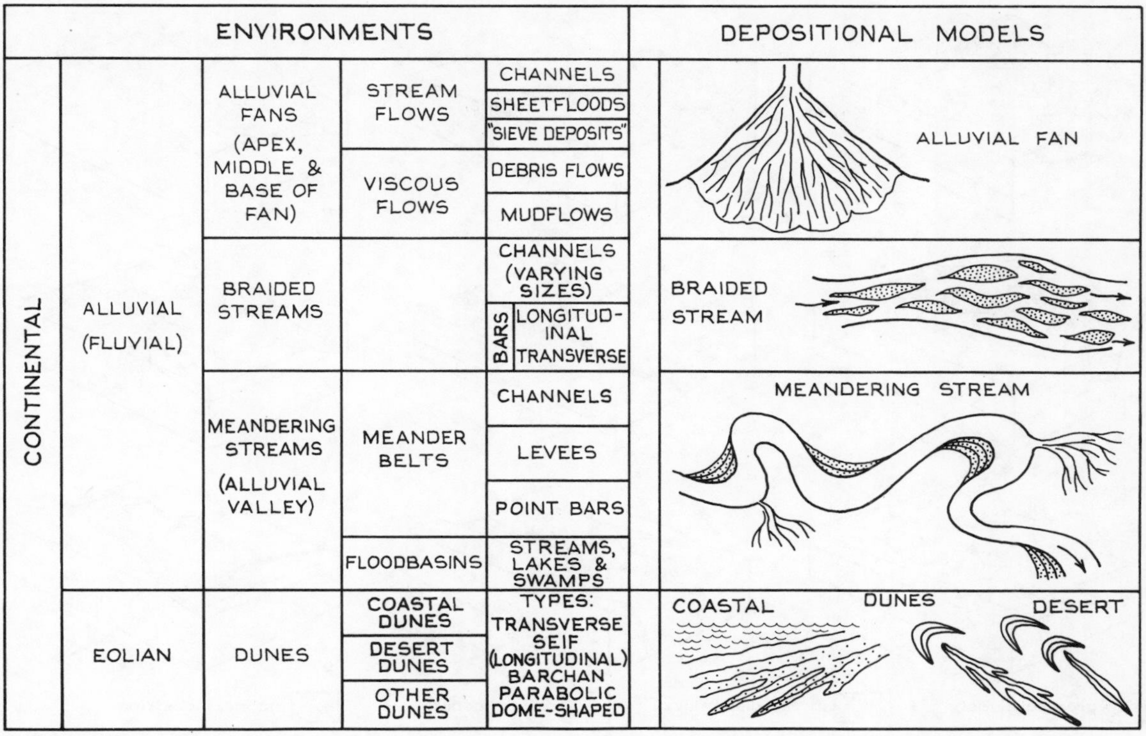

ENVIRONMENTS				DEPOSITIONAL MODELS	
CONTINENTAL	ALLUVIAL (FLUVIAL)	ALLUVIAL FANS (APEX, MIDDLE & BASE OF FAN)	STREAM FLOWS	CHANNELS	ALLUVIAL FAN
				SHEETFLOODS	
				"SIEVE DEPOSITS"	
			VISCOUS FLOWS	DEBRIS FLOWS	
				MUDFLOWS	
		BRAIDED STREAMS		CHANNELS (VARYING SIZES)	BRAIDED STREAM
				BARS LONGITUDINAL	
				TRANSVERSE	
		MEANDERING STREAMS (ALLUVIAL VALLEY)	MEANDER BELTS	CHANNELS	MEANDERING STREAM
				LEVEES	
				POINT BARS	
			FLOODBASINS	STREAMS, LAKES & SWAMPS	
	EOLIAN	DUNES	COASTAL DUNES	TYPES: TRANSVERSE SEIF (LONGITUDINAL) BARCHAN PARABOLIC DOME-SHAPED	COASTAL DUNES DESERT
			DESERT DUNES		
			OTHER DUNES		

Figure 6.2. Continental sedimentation occurs in a variety of different environments within which the depositional models are relatively limited. These models appear, however, within a wide range of tectonic and topographic settings as shown in figure 6.3 (after Conybeare, 1979).

may be limited in number but distinctive for the particular facies. Rapid lateral and vertical variation is common but in many successions the range of different lithologies is actually rather small. Trace fossils may be present, locally abundant but few in kind. Plant and vertebrate fossils tend to be thanatocoenose disarticulated parts or fragments. In general these fossils are rare and are limited to relatively few beds within most sequences. Entire organisms were more commonly preserved in lacustrine formations than elsewhere, although at certain levels they may be locally abundant.

Several depositional facies may be distinguished (Allen, 1979):

Aeolian. Well sorted, medium-grained sandstones in cross-bedded sets of 5 m or so thick. Sets are divided from each other by flat to curved erosion surfaces. Facies interfingers with alluvium.

Proximal alluvial. Sheet-like or channel-fill bodies of moderately to well sorted and locally cross-bedded or cobble conglomerate (stream deposits), poorly sorted and bedded, matrix-supported conglomerates and breccio-conglomerates (mudflow deposits); coarse mudstones with sandstone partings, locally calcretised. Some trace fossils.

Medial alluvial. Well-bedded, relatively coarse-grained sandstones with minor exotic, mixed—or intraclast—conglomerates; mudstone with sandstone interbeds; calcretes. Trace fossils varied, plant remains and vertebrates.

Distal alluvial. Mudstones equal or exceed in volume the sandstones; poorly bedded, thick, carbonate palaeosols present. Intraformational conglomerates with plant and vertebrate fossils.

Lacustrine shallow water. Coarse to fine sandstones and mudstones in varying proportions, well

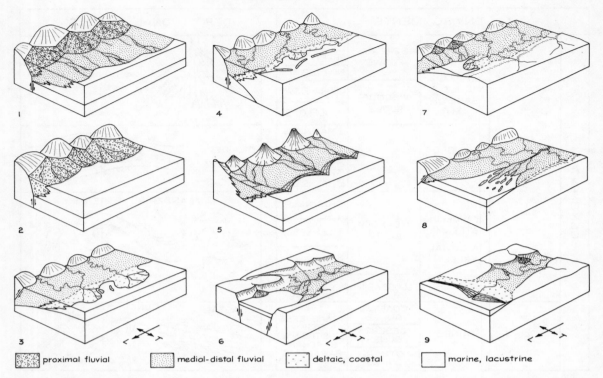

proximal fluvial   medial-distal fluvial   deltaic, coastal   marine, lacustrine

Figure 6.3. Basin-fill patterns as modelled by Miall (1981) and widely applicable to Devonian examples. L =longitudinal, T =transverse. Scales are variable and most basin-fill patterns can occur in several tectonic settings. Examples: 1, Lower Devonian, Anglo-Welsh Cuvette; 2, Scarclet Group, north-east Scotland; 3, Hecla and Weatherall Formations, Arctic Canada; 4, Peel Sound Formation, Arctic Canada; 5, Lower Devonian Midland Valley of Scotland; 6, Cannes de Roche Formation, eastern Canada; 7, Lower Devonian, Anglo-Welsh Cuvette, and Catskill Group, eastern U.S.A.; 8, Grey Hoek Formation and Wood Bay Series, Spitsbergen, and Caithness Flagstone Group, Scotland; 9, Wijde Bay Group, Spitsbergen.

bedded or laminated, thin; many sedimentary structures, and rare pseudomorphs after evaporites; plant, vertebrate and invertebrate fossils.

Lacustrine deep water. Sandstones and shales or mudstones, thin to massively bedded, may be resedimented; pyrite common. Plant fragments common in some deposits, other fossils may be highly localised or stratigraphically confined.

Coastal mudflat. Cross-bedded sandstones with mudstones and thinly bedded sandstones in fining-upwards sequences, mud cracks, calcretes, erosion surfaces. Trace fossils common. Plant remains and invertebrates (molluscs, ostracodes, lingulids) and some vertebrates disarticulated or in place.

Coastal barrier. Coarsening-upwards sequences of mudstones and sandstones. Wide variety of sedimentary (bedding) structures, trace fossils and

marine fauna (lingulids, bivalves, gastropods, ostracods).

Inshore marine. Mudstones with few or no sandstones, with abundant and diverse marine fauna including calcareous brachiopods, corals and crinoids.

The last of these facies can scarcely be regarded as continental, but it occurs interfingering with those that are, and it may be red, buff or grey-green in colour.

Within the Devonian system many of these environments are represented by contiguous facies so that laterally one may pass from the alluvial into the lacustrine, deltaic or marine. Such lateral changes occur in many parts of the world; in fact wherever Old Red Sandstone facies occur in external basins there are lateral changes of a major kind. Rapid accumulation of a continental molasse at the border of a sea seems to have been widespread and to have produced local marine regression. Fault activity continuously influenced the development of many of the external basins, and in some internal basins it even reversed the palaeo-slope of the basin axis.

In many parts of the world (North America, Europe, China, Australia, Antarctica) Devonian palaeosols are now recognised. These distinctive deposits are most commonly carbonates (calcretes); some are siliceous and they appear to require several thousand years for their formation in areas beyond the active floodplains (Allen, 1974). They may have been influenced by terrestrial vegetation and are thought to form today only in areas where ground water is drawn by capillary action upwards through the regolith. They are laterally extensive and continuous to such a degree that in the Lower Old Red Sandstone of Britain palaeosols are sufficiently widespread to be of use in correlations.

Friend (1978) examined the distinctive features of certain Devonian and Tertiary alluvial deposits, noting that in some respects the ancient river systems appear to differ from those of the present day. For example, Old Red Sandstone sequences and structures suggest that the sediment accumulated in areas of down-stream decrease in river size, with water evaporating and entering the substrate at rapid rates. Streams did not cut deeply into the regolith over much of their length but the alluvium in some areas was built up into fans or lobes with convex upper surfaces. The lack of vegetation over perhaps most of the drainage slopes and the proximity of continuing orogenic activity were important controls of the ancient fluvial systems.

Allen (1964, 1965, 1974, 1977, 1979) used the term *cyclothem* to describe the repeated upward-fining sandstone and siltstone units of the Old Red Sandstone. Cyclic sedimentation of another kind occurs in the lacustrine facies of the Middle Devonian of northern Scotland. Here climatically controlled cycles resulting from the rise and fall of lake levels in the Orcadian Basin have been described by R. N. Donovan and others (Donovan, R. N. *et al.*, 1974, 1976; Donovan, R. N., 1975). The cycles are revealed in four lithological associations, each being made up of drab, thinly bedded mud-

stones, siltstones (some calcareous) and minor amounts of carbon, sulphides and massive fine sandstone. Sedimentary structures abound and fish remains are abundant at repeated levels. R. N. Donovan shows that these beds could have originated in lakes that varied from deep (stratified) to shallow and impermanent. The lakes within the Orcadian Basin had an estimated area of up to 50 000 km² and their dimensions varied with the climate, as do those of African lakes today.

Coarsening-upward cycles have been described from the Devonian Hornelen Basin of western Norway (Steel, 1976; Steel *et al.*, 1977; Steel and Aasheim, 1978). Here some 25 km of sediments are organised into more than 150 basin-wide cycles, each about 100 m thick and consisting of marginally derived fanglomerates and laterally equivalent alluvial-plain sediments. These cycles are believed to reflect a response to the tectonic lowering of the basin floor (that is, to the prograding of talus and alluvial fans). Repetitive climatic variations may, however, be invoked. From time to time there was a progressive shift of the locus of subsidence by about 0.25 km. Possibly this shift is related to discontinuous strike–slip along a major local wrench fault, a good example of tectonic control of basin sedimentation. On Bjørnøya, also, coarsening-upward sequences have been held as indicative of delta-like progradation into quiet water (Gjelberg, 1977).

Glacial or periglacial deposits of Devonian age have so far only been reported from South America, and they are difficult to interpret in detail. Nevertheless the tillites, varied sediments and other features recorded in north-eastern Brazil (Maack, 1957, 1960; Malzahn, 1957) indicate an undoubtedly cold climate.

The Devonian Continental Biota

The flora

Records of pre-Devonian continental flora are relatively few and there is some doubt as to the age of the oldest undoubted terrestrial plants. The bryophytes (mosses and liverworts) need moist environments for survival, and may have been well established before the beginning of the period. The tracheophytes, with their vascular systems, well-differentiated stem systems, evolved to meet the demands of the drier environment. Devonian tracheophytes were relatively simple (forms) but adapted for growth in the open air, a step which allowed plant life to advance rapidly during the period. By mid-Devonian time much of the land in moist climatic areas must have supported a simple but expanding flora. From the narrow (psilophyte flora) 'sedge-meadows' of Early Devonian days a change had been made to a thicket or even forest-like vegetation that was several metres high. At the close of the period *Archaeopteris* forests were widely established with even higher vegetation and with an 'undergrowth' and newly developed ecosystems that involved animals as well as plants. The Psilopsida appear to have given rise to the Lycopsida, Sphenopsida and Pteropsida early in the Devonian and even the Pteridospermopsida had evolved by the end of the period. Reproduction by means of spores was in this group replaced by seeds, and confinement to the wet environments was thus no longer necessary. The geographical spread of plants was thereafter, no doubt, rapid and with immediate ecological (and geological) consequences. Whereas at the end of the Silurian *Cooksonia* and other plants were only a few centimetres high, *Archaeopteris* had a trunk diameter of over a metre and was presumably a large tree. As Chaloner and Sheerin (1979, p. 152) point out, this one simple feature represents the increasing potential of plants to 'form more complex communities, in which ecological 'stratification' of the constituents would evolve, with different micro-environments at different levels of light intensity and humidity'.

Putting together the fossil evidence for this model is fraught with difficulties. Many occurrences of fossil plants in these continental rocks are at local, isolated single exposures where the stratigraphic position may not be clear. The fossils are commonly very fragmented, being poorly preserved in coarse sediment; most genera are based upon single detached items—leaves, axes, sporangia, etc. Different parts of the same plant have been identified as separate taxa. For example, the name *Callixylon* has been given to pre-mineralised woody stems and *Archaeopteris* to leaves and shoots of the same plant (Beck, 1960).

Early accounts of the Devonian flora were by Miller (1859) in Scotland and Dawson (1862) on Canadian material and, although limited, these observations established that the Early Devonian flora was widespread and varied. A hundred years later it was apparent that there were distinctive floras for each Devonian epoch. *Psilophyton, Hyenia* and *Archaeopteris* respectively were generally regarded as broad stratigraphic markers for each Devonian series. It is striking that the vascular plants underwent an evolutionary explosion in Siegenian time. The record from this stage very obviously includes a large number of new taxa, on a wider scale, than in any previous stage. The actual origins of the many new genera in the Siegenian are not beyond question and the Early Devonian plant communities were simple in composition. Devonian macroplant fossils have now been recorded from all parts of the North Atlantic area, western and northern North America, many parts of the U.S.S.R. and China, Australia and Antarctica. The earlier Devonian floras seem to be very uniform but by Late Devonian time there may have been the beginnings of the distinct provinciality that was to reach its fullest development later in the Palaeozoic. It is now clear that three major world floras are known: *Zosterophyllum* = Lower Devonian; *Protolepidodendron* = Middle Devonian; *Leptophloem rhombicum* = Upper Devonian. While local correlation by spores seems soon to be possible the use of macroplant fossils in such a role is less certain. However Banks (1980) has proposed seven assemblage zones for the span from Pridolian (Silurian) to post-Famennian (see figure 6.4).

It could be argued that the surge of photosynthetic activity by the green terrestrial flora of Devonian times increased the proportion of atmospheric oxygen. This in time may have enhanced vertebrate activity and evolution.

Invertebrates

The colonisation of fresh waters and the land surface by plants opened the way for an expansion of the animal phyla that began in pre-Devonian times. During the Devonian period, however, this expansion must have gathered momentum (see Rolfe, 1980). The new terrestrial environments were seemingly at first occupied by few predators and were used even by animals of aquatic habit early in

Series or Stage	Banks, 1980	McGregor, 1977		Richardson 1974
Post Famennian Tn Ib Tn Ia	_ _ _ ? _ _ _			V. nitidus
FAMENNIAN	Rhacophyton Assemblage-zone VII			V. pusillites S. lepidophytus
				L. cristifer
FRASNIAN	Archaeopteris Assemblage-zone VI			optivus-bullatus
GIVETIAN	Svalbardia Assemblage-zone V			Triangulatus
		devonicus-orcadensis		Densosporites devonicus
EIFELIAN	Hyenia Assemblage-zone IV	velata-langii		Rhabdosporites langii Acinosporites acanthomammillatus
UPPER		annulatus-lindlarensis	Grandispora	Calyptosporites biornatus-proteus
EMSIAN	Psilophyton		sextantii	Emphanisporites annulatus
LOWER	Assemblage-zone III	caperatus-emsiensis		
SIEGENIAN				Dibolisporites cf. gibberosus
	Zosterophyllum			Emphanisporites micrornatus
GEDINNIAN	Assemblage-zone II	micrornatus-proteus		Streelispora newportensis
PRIDOLIAN	Cooksonia Assemblage-zone I	chulus-?vermiculata		Synorisporites tripapillatus

Figure 6.4. Comparison of the generic assemblage zones recognised by H. W. Banks with the palynological zones of McGregor (1977) and Richardson (1974). The latter's scheme has already been referred to in figure 5.14. The broader divisions of the system seem to be well defined by plant assemblage zones and spore zones over wide areas of the earth, but most occurrences are within what must have been the warmer realms (see page 81).

life to escape predators later in their life histories. The metamorphoses of the insect world were to be rapidly evolved. Complex terrestrial ecosystems were in existence by the beginning of Carboniferous time. Fossil evidence of the range and relationships of all these organisms is, however, not unexpectedly rare. Aquatic habitats were undoubtedly the more fully occupied early in the period and fossils from these are relatively better known than those of land animals. It was a time of explosive evolution amongst marine bivalvia and may have been so also for the non-marine forms. The oldest fresh-water unionid genus, *Archanodon*, is common as are representatives of the Grammysiidae, Modiomorphidae etc. Commonly found in the marine–

non-marine transitional facies with bivalvia are gastropods, but so far they do not appear to have penetrated into the entirely non-marine or terrestrial Devonian realms. Those that do occur (*Platyschisma* etc.) were probably epifaunal algal grazers.

The arthropoda are represented by a wide range of fossils, though rare—merostomata, crustacea, mites and spiders, and miriapods. The crustacean ostracods include a very large number of families, almost all of which are marine. Large leperditiids and others are known from non-marine facies in Europe and North America, where they are common at a few levels and rare elsewhere.

The acarid *Protocarus crani* from the Rhynie

Chert is one of the oldest recorded arachnids, but Petrunkevitch (1953) believed that all the arachnid orders were already extant (or in the process of evolving) in Silurian time. Spectacular proportions as well as considerable taxonomic diversity were achieved by the non-marine merostomata in the shape of the eurypterids. The Devonian *Pterygotus* (*Erettopterus*) was the largest arthropod of all time, reaching nearly 2 m in length. Xiphosurans are known from the continental Upper Devonian of Pennsylvania. Conchostracans locally appear to have inhabited marginal marine sites and lacustrine environments in Eurasia and Antarctica.

Trace fossils are reported from various facies of the Old Red Sandstone and the marine–non-marine transitional facies. They include crawling and resting traces (Trewin, 1976; Allen and Williams, 1981a, 1981b), possible feeding traces and dwelling structures (Narbonne *et al.*, 1979). *Beaconites* (= *Laminites*), a very large meniscus-filled burrow known in Antarctica and in Great Britain, possibly housed a large but unknown invertebrate.

Vertebrate faunas

Vertebrate remains found in several continental facies of the Devonian seemingly represent animals that are fitted to a wide variety of habitats — lacustrine, fluvial, tidal and swamp. In some lakes the faunas were clearly large and diverse. While many occurrences are biocoenoses, the majority are thanatocoenose accumulations of disarticulated and transported remains.

The fossils consist of bones, teeth, scales and the armour of head and thoracic regions. Much of the armour is highly ornamented but this may not be of great use in taxonomy. As with the floras, many vertebrate stocks show not only an increase in number and diversity of taxa with time but also an increase in body size. Details of the anatomy of many forms are still unknown and the modes of life and ecology of the majority of species is conjectural to some degree.

The agnatha are the most numerous group in Lower Old Red Sandstone deposits throughout the world and the faunas include as many as fifteen taxa, commonly with acanthodians and other placoderms present as well. Of all the agnatha the thelodonti may be the most widespread and cosmopolitan (Turner, S. and Tarling, 1982). They probably were able to spend periods of time in the sea and to migrate relatively rapidly and far. Three general realms or provinces may be distinguished in the Lower Devonian — North Atlantic, Siberian and Chinese — based upon continental areas between which migration was comparatively difficult. Middle Devonian agnatha are rare and in the Upper Devonian only four or five genera remain. Their extinction seems to have been perhaps completed at the Famennian regression.

The antiarchi were well established in China by the beginning of Devonian time (P'an, 1980) but reached other areas only in the Middle Devonian. The later Middle and Upper Devonian antiarch faunas appear to be generally similar everywhere, although a few endemic east-Asian genera remain.

The arthrodires were common in Early Devonian habitats everywhere except in China, although they were well established even there in Middle and Late Devonian times. Other placoderms were widely distributed from Greenland to Antarctica especially in the upper two Devonian series.

The bony fishes are found in both marine and continental facies but only the crossopterygians and dipnoi are present in the Lower Devonian; some forms in later deposits are very widespread and there was clearly a simple broad equatorial(?) realm that all these groups occupied.

Details of stratigraphic distribution are given by Westoll (1979) who demonstrates that most placoderms had died out well before the end of the period while perhaps the Onychodontida alone of the jawed fishes had become extinct by the beginning of the Carboniferous.

Figure 6.5 shows the palaeontological 'zones' and assemblages in Old Red Sandstone formations of the British Isles.

Distribution of Old Red Sandstone Facies

Within and around the Devonian orogenic belts thick red beds accumulated under a wide range of conditions. As post-orogenic molasse, and as widespread reworked regolith over extensive areas of the continents, they were a conspicuous feature of Devonian continental and transitional deposition

UPPER	**SCOTLAND**	**Assemblages**	*Bothriolepis cristata* *Phyllolepis concentrica, Glyptopomus minor* *Bothriolepis major* *Asterolepis maxima*		
MIDDLE			*Watsonosteus fletti* *Millerosteus minor* *Asmussia murchisoniana* *Dickosteus threiplandi* *Palaeospondylus gunni* *Coccosteus cuspidatus* *Thursius macrolepidotus*		
LOWER	**Anglo-Welsh**	**Borderlands 'Zones'**	*Rhinopteraspis dunensis* *Pteraspis (Belgicaspis) crouchi* *Protopteraspis leathensis* *Traquairaspis symmondsi* *Traquairaspis pococki* *Hemicyclaspis*	*Turinia pagei* assemblage *Gonioporus/Katoporus* assemblages *Thelodus parvidens* assemblage	

Figure 6.5. Palaeontological 'zones' and assemblages in the Old Red Sandstone formations of the British Isles. Thelodont assemblages are shown at lower right for the Lower Old Redstones of the Welsh borderlands (based on works by E. I. White, T. S. Westoll, S. Turner and others). There is a particular dearth of palaeontological data by which to define the Lower/Middle and Middle/Upper Old Red Sandstone boundaries.

in the North Atlantic area (see figure 6.6) and the Canadian Arctic, Siberia, China, Australia and Antarctica.

The *North Atlantic area* has been the most intensively studied, the once-contiguous areas of continental sedimentation being now, by erosion, tectonics and continental rifting, scattered over great distances (Friend, 1968: Allen *et al.*, 1968). The grand event that initiated Old Red Sandstone deposition was the closure of the Iapetus Ocean in the Late Silurian–Early Devonian. Late Silurian red beds mark the onset of continental environments; subsequent basal red bed formations are almost all unconformable upon eroded, deformed and metamorphosed Lower Palaeozoic rocks. In the Appalachian extension of the Caledonides many red beds owe their origin to later (Acadian) earth movements. A link between the Appalachian orogenic belt and the Cordilleran via the southern-most U.S.A. is hypothetical but in the west red beds are associated with the Antler and other orogenic phases (see below).

The outcrops extend in a belt from Spitsbergen (76–81°N) (Friend, 1961, 1969, 1981) to the northeastern U.S.A. (Schluger, 1976a, 1976b), a distance of over 2500 km. Leeder (1976) and Greiner (1978) have stressed the close relationship of Devonian sedimentary facies to the tectonics of the Caledonide belt and the northern margin of the supposed Hercynian ocean, with fault control of basin subsidence being paramount during Devonian and Carboniferous times. Deposition occurred early in local basins, many of which were defined by these NE–SW-trending faults, or by hinge lines in the underlying basement or by simple downwarp. Later sedimentation masked some of these marginal features and spread from one basin to another. Within these basins synorogenic to post-orogenic fluvial molasse and lacustrine deposits reached thicknesses of 6000 m or more and were up to seven times thicker than the sediment that spilled eventually on to the adjacent blocks (Leeder, 1976). Local sedimentation was extremely rapid and accompanied by calc-alkaline volcanicity in Scotland

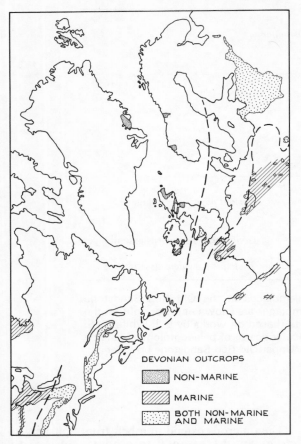

DEVONIAN OUTCROPS

NON-MARINE

MARINE

BOTH NON-MARINE AND MARINE

Figure 6.6. The main Devonian outcrops in the North Atlantic region (after Friend, 1969). Marine Devonian rocks are also known at depth in Southern England and in the northern North Sea, and many small outcrops of Devonian also occur in eastern-most Canada (including Newfoundland). The Caledonide ranges largely enclosed the non-marine basins (west of dashed line) and the Central European sea may never have receded beyond the eastern dashed line.

and elsewhere. Tectonic activity continued in innumerable spasms throughout Early and Late Devonian times, being especially vigorous in the Middle Devonian. This middle phase corresponded to one of widespread transgression, subsidence and slow pelagic sedimentation in the south-west of England (Webby, 1965).

Woodrow *et al.* (1973) have argued for the origin of the North Atlantic Old Red Sandstone under hot or semi-arid climate in an equatorial belt and under a more even pattern of greater rainfall to the north

where the coals of northern Canada, Spitsbergen and Bear Island were laid down (see figure 6.7).

In Spitsbergen the northern-most outcrops of Atlantic Old Red Sandstone have been closely studied by British and Polish geologists; French, Norwegian and Swedish palaeontologists have been active in describing the remarkable Lower Devonian vertebrate faunas. The Siluro–Devonian rocks of Vestspitsbergen rest with a sharp unconformity upon the metamorphic Hecla Hoek Series. The main outcrop is in north Vestspitsbergen but a small area of Devonian also occurs near Hornsund in the south. All were folded and faulted during an Upper Devonian (Svalbardian) orogenic phase. The formations range in age from Upper Downtonian (Pridoli–Lower Gedinnian) to upper Middle Devonian (Upper Givetian) or lowermost Upper Devonian (Lower Frasnian) (figure 6.8). Environments of deposition included torrential alluvial fans, river flood plains, inland lakes and brackish lagoons. There are plants, invertebrates, vertebrates and trace fossils locally in abundance. Most famous are the plants (Hoeg, 1942) and the vertebrate faunas which are largely thanatocoenoses of river-dwelling or lake-dwelling forms. Cyathaspidids, pteraspidids, arthrodires, antiarchs and others show strong similarities to the faunas of Europe and eastern North America but there are many genera known so far only from Spitsbergen (for lists see Friend, 1961).

Upper Devonian rocks on Bear Island (Bjørnøya) comprise a thick series of sandstones with minor shale and coal bands (Gjelberg, 1977) reaching a thickness of 600 m. The flora is rich in *Archaeopteris* and *Bothrodendron* and two facies— flood plain and lacustrine–deltaic—are distinguished.

In east Greenland intensive studies of the Old Red Sandstone were recently completed (Friend *et al.*, 1976a, 1976b; Yeats and Friend, 1978), using a methodology that allowed rapid and detailed sedimentary logging on computer-readable forms. Particular attention was paid to sedimentary structures and to the occurrence of the fossils. The vertebrates occur mostly as disarticulated bone thanatocoenoses while the plant fossils, up to 2 m in length, appear to have been transported. All of this evidence has been used to model the alluvial and lacustrine paleoenvironments and the structural

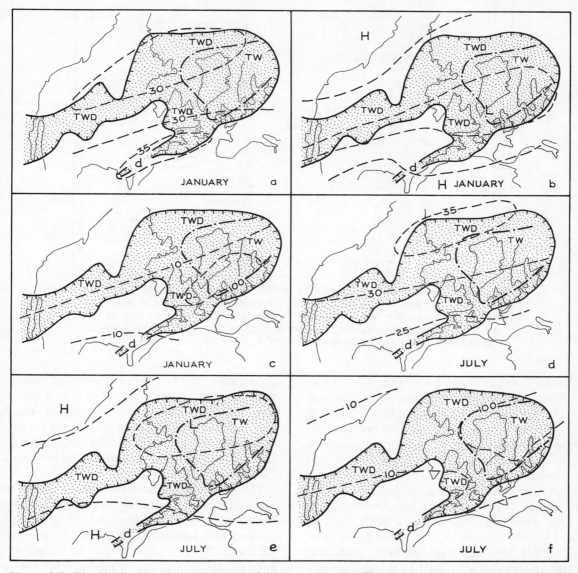

Figure 6.7. The North Atlantic land mass and its extension, the 'Transcontinental Arch', across North America have been thought by Woodrow *et al.* (1973) to possess the climatic regimes shown here. Suggested temperatures are in °C and precipitation is in centimetres. Climatic zones are shown by the letters TW (tropical wet), TWD (tropical wet and dry), and d (desert).

relationships of the formations have been studied in an attempt to unravel the sequence of contemporaneous tectonic events. Deposition, following on the erosion of Caledonian metamorphosed and deformed rocks, was preceded by the extrusion of volcanics. Some 1500 m of upper Middle Devonian clastic rocks accumulated, essentially from nearby source areas. This was followed by a further 2500 m of material from an easterly source and a shift in sedimentation westwards with faulting, local folding and erosion. Further faulting occurred prior to the deposition of more than 3300 m of Upper

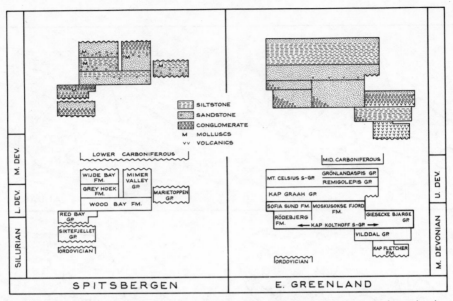

Figure 6.8. The broad lithostratigraphy of the Old Red Sandstone deposits in Spitsbergen and east Greenland illustrates the spread of deposition in small post-orogenic (fault) basins that are floored by metamorphosed Ordovician or older rocks. The Spitsbergen deposits reach an aggregate thickness of about 6500 m, while those in Greenland perhaps attain 7000 m (after Friend, 1973).

Devonian. During this epoch the sedimentary basin may have been widened with fluvial, lacustrine and aeolian deposition and an abundant supply of material from the east. The marginal breccias and conglomerates give way in a short distance to alluvial or lacustrine sandstones and siltstones, and the migratory nature of the lakes and rivers within this active basin is most apparent.

In western Norway the Old Red Sandstone has received much attention in recent years (Nilson, 1968, 1973; Brynhi and Skjerlie, 1975; Steel, 1976; Brynhi, 1978; Steel and Aasheim, 1978); particular interest here has also attached to the sedimentary response to continuing tectonism. The present Old Red Sandstone outcrops are small and scattered but they reach a thickness of about 25 000 m and indicate several small basins that are filled with 'cyclic' deposits. Coarse and very coarse sediments dominate what are clearly intramontane facies of Lower and Middle Devonian age. Viscous mud-flows, flood sheets and other local sedimentary bodies line the unconformable base of the basins in western Norway. Local relief was high and the source of most sedimentary material was essen-

tially local. Steel (1976) has shown that successive pulses of very coarse clastic debris were a repetitive response to continuing fault movement, regional uplift and the rapid lowering of the basin floors. An extramontane facies in the Ringerike Series of Siluro–Devonian tidal flat and alluvial sediments in the Oslo area is also known.

The Orcadian–Shetland province of the Scottish Old Red Sandstone extends over a large part of the northern North Sea (see figure 6.9). Westoll and co-workers (see Donovan, R. N. and Foster, 1972; Donovan, R. N. *et al.*, 1974, 1976; House, 1977) have extensively studied the marginal breccias and alluvial sandstones and the lacustrine 'Caithness Flagstone Series'. 8000 m or more of sedimentary infill remain in the shallow but subsiding basin which was perhaps as large as 50 000 km^2 and occupied by shifting and impermanent bodies of water, some of which were hypersaline from time to time. Lacustrine transgression and regression took place and signs of strong evaporation are present. Much of the succession is a laminated carbonaceous–calcareous siltstone facies, deposited in tropical, stratified lakes, which perhaps sup-

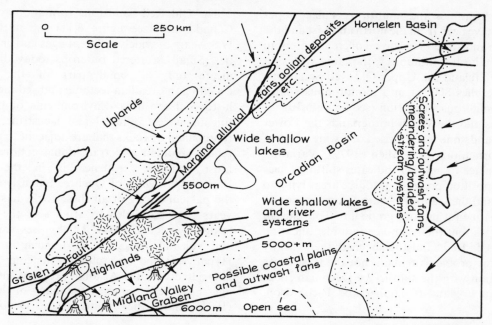

Figure 6.9. The Orcadian basin contained over 5000 m of non-marine clastic deposits, largely of Middle Devonian age. At the northern end of the marine North Sea Basin, its surface was occupied by environments of alluvial lacustrine and aeolian deposition. Subsidence was irregular and influenced by large faults: the provenance of sediment was from the Caledonian–Fenno–Scandian uplands.

ported an extensive biota (Donovan, R. N., 1980). Stromatolites were locally abundant at the margins. Some Lower Old Red Sandstone is present but most formations are of Middle Old Red Sandstone age. Overlying the 'Caithness Flagstones' are Upper Devonian sandstones and other clastics, with lavas in Orkney, as much as 1760 m thick. On the mainland they are alluvial deposits that were carried from the highlands to the west or south-west. The smaller south Shetland Basin was perhaps connected to the Orcadian and included aeolian deposits.

The 'Caledonian Cuvette' of the Midland Valley of Scotland shows Upper Old Red Sandstone that rests upon Lower Old Red Sandstone or older formations and grades up into Carboniferous strata. It presents a complex lithostratigraphy and has long been known for its vertebrate and other fossils (see Westoll, 1977). Sedimentological studies have shown the importance of locally derived coarse clastic materials in scree and fan deposits, and braided and meandering streams. Calcretes are widespread. Volcanic rocks, both andesitic and basaltic, are present in great thickness. The depositional environments of the sediments and the occurrence of the vulcanicity are related to the activity of the bounding faults of this basin (Morton, 1979). Deposition did, however, in Late Old Red Sandstone time extend beyond the rifted zone to the Cheviot area and perhaps to the north of Ireland. In the Cheviots volcanic rocks, thick alluvial clastic deposits and palaeosol accumulations indicate rapid deposition on a volcanic and irregular upland topography.

The Old Red Sandstone of Ireland (see Holland 1977, 1979; Clayton *et al.*, 1980) is scattered in outcrops between the Fintona area in the north and the south coast. The stratigraphy is complex, the formations generally being unconformable upon Lower Palaeozoics. Most outcrops show Upper Old Red Sandstone, but palaeontological evidence is patchy. Local unconformities and facies changes occur and it is clear that alluvial and lacustrine environments in Ireland were influenced by contemporary tectonics and landforms. Palaeoslopes appear to have been from the north and east and the Old Red Sandstone may have covered the entire country and been continuous with that in Wales. In south-west Ireland the Upper Old Red Sandstone possibly reaches 8000 m in a monotonous coarse and unfossiliferous succession (see figure 6.10).

In Wales and the Welsh borderlands the Lower Old Red Sandstone is comprised of a wide range of alluvial formations (see Allen, 1979). Variation from one place to another indicates shifting facies belts as deposition proceeded, source areas lying to the north and the sea to the south. Thin Upper Old Red Sandstone formations grade into Tournaisian marine strata, but there is no certain identification of Middle Old Red Sandstone beneath them. Many stratigraphic problems remain, not the least those of correlation with the subsurface Devonian rocks of southern England (Chaloner and Richardson, 1977).

The Old Red Sandstone formations of eastern Canada and (see figure 6.11) the north-eastern U.S.A. are of widely different ages and characteristics. Small, scattered outcrops today directly indicate only a small part of the complex Caledonian–Acadian tectonic and sedimentational history of the area. Environments of truly continental aspect, alluvial and lacustrine, are here represented by rocks that are adjacent to those that reflect lagoonal and transitional conditions. The most complete sequence is in the Chaleur Bay–Gaspé region of Quebec. Extensive work on the palaeontology of all these rocks has in recent years been paralleled by studies of the sedimentology (see Rust, 1981). Thicknesses vary greatly and volcanic rocks are also much in evidence. Hypabyssal and also plutonic igneous rocks are associated with the Acadian orogeny and isotopic ages have been obtained for many of them. Rust (1981) relates the origins of the Devonian and

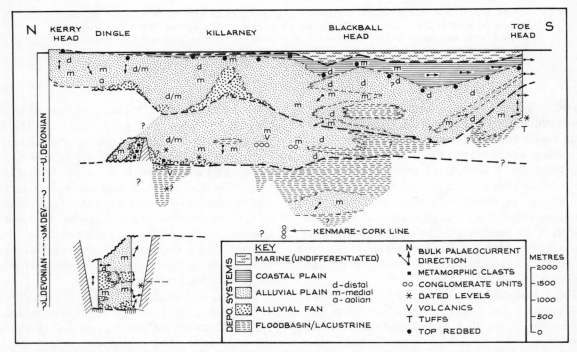

Figure 6.10. Gardiner and MacCarthy's (1981) interpretation of the Old Red Sandstone of the (external) Dingle–Shannon and Munster basins, south-west Eire, postulates a succession of *depositional episodes* that span perhaps the entire period. During each episode a distinct *depositional system*, a facies assemblage that represents major physiographic units, was developed. This section across south-west Eire shows episodes 1A to 5, each episode being marked by a hiatus or change of system as the basin developed. Local changes in rates of subsidence and sedimentation probably depended upon the behaviour of major inherited lineaments within the Caledonian orogenic belt.

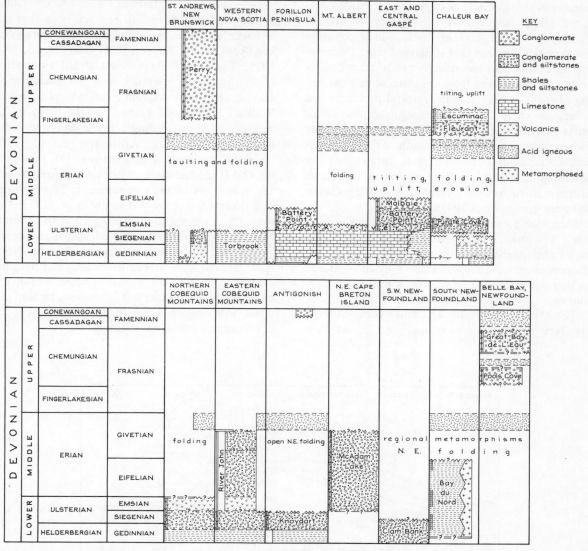

Figure 6.11. Correlation of the Devonian formations of eastern Canada (after Poole *et al.*, 1970). The early clastic phases of deposition, the widespread acidic igneous activity and uplift (Acadian Orogeny) and the more restricted later Devonian clastic sedimentation are well distinguished here.

Carboniferous continental rocks of eastern Gaspé to broad upwarping that results from the activity of a shallow subduction zone on the nearby continental margin, activity being at its maximum in mid-Devonian Acadian Orogeny.

Most widespread are the Lower Old Red Sandstone formations. In Nova Scotia the Knoydart Formation closely resembles the Ditton series of South Wales (Boucot *et al.*, 1974) and rests conform-

ably upon transitional to marine facies. Coarse clastic formations in Newfoundland overlie with discordance strongly deformed rocks. 500–2000 m of Middle Devonian fluviatile rocks occur in northern Nova Scotia.

In northern Maine, U.S.A., and New Brunswick, Canada, a Late Devonian phase of fault–trough deposition involved alluvial boulder fans, fluvial gravels, sands and lacustrine silts or muds, with numerous thin calcareous duricrusts or palaeosols. An example is the Perry Formation in Maine (Schluger, 1976a), 1000 m thick, which contains primary red pigment that is derived from underlying rocks and also secondarily produced red minerals that are derived from the syngenetic weathering of mafic minerals by intrastratal movement of solutions. There is evidence of saline deposition in parts of the lacustrine facies. Tectonic activity was recurrent in eastern Quebec; south of Gaspé extensive continental deposition was interrupted by several phases of both faulting and uplift. Large plutonic intrusives were emplaced, and the Gaspé Sandstone molasse is 300 m thick.

Devonian formations in Newfoundland include thick volcanics, mixed marine and non-marine facies and coarsely clastic non-marine beds. Offshore occurrences may be increasingly encountered as petroleum drilling continues; some are already known.

The Middle and Upper Devonian rocks of western New York State (see figure 6.12) and Pennsylvania include the famous Catskill facies, 1200 m or more of alluvial and deltaic red beds that thin westwards and interfinger with marine strata (Walker, R. G., 1971; Walker, R. G. and Harms, 1971, 1975). The lower 600 m of the Catskill 'Formation' in Pennsylvania contains over 20 marine–nonmarine regressive sequences and is regarded by Walker, R. G. and Harms (1971, 1975) as the deposits of a prograding muddy shoreline rather than a typical delta. Allen and Friend (1968) recorded many similarities between this facies and the Old Red Sandstone of Britain and Spitsbergen. Their analysis of the environments of deposition was taken further by Woodrow *et al.* (1973) who went on to suggest the general setting and climatic regimes under which the alluvial, intermontane and coastal plain deposits were developed on the North Atlantic land mass. In particular they distinguished evidence for different fluvial systems, braided or meandering, carbonate soil formation, rainfall patterns, evaporation rates, wind patterns and the development of coastal swamps.

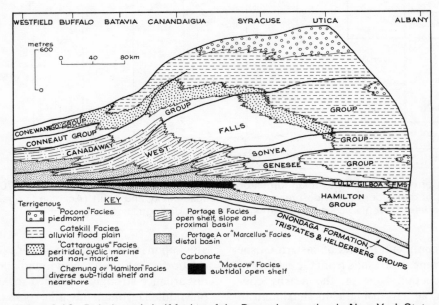

Figure 6.12. Deltaic and shelf facies of the Devonian section in New York State. The continental to marine transition is now well documented from both the palaeontological and sedimentological points of view but fossils are relatively rare within the Catskill and Pocono facies. Palaeocurrent directions are westerly from the 'Acadian' highlands into the Appalachian basin and over the epicontinental seafloor to the west.

Arctic Canada

The Canadian Arctic Archipelago offers two remarkable developments of Devonian continental rocks. They are the Middle Devonian clastic wedge spread over some 200 000 km² of Banks, Prince Patrick, Melville and Bathurst Islands (Embry and Klovan, 1976), and the Lower Devonian Peel Sound Formation of Prince of Wales and Somerset Islands (Miall, 1970a, 1970b). In overall aspect and proportions the clastic wedge is miogeosynclinal and indicative of major crustal subsidence. It is comprised of about 5000 m of formations which

represent facies that range from alluvial plain with meandering, and braided, streams to submarine fan and open marine shelf. Source areas to the north and east include the Caledonian and Pearya Mountain systems and the Precambrian of Greenland (see figure 6.13). The fluvial strata are thought to have been deposited under tropical savannah to humid or very humid conditions. Deposition began in the extreme north-east in Eifelian time and a coastal plain prograded south and west to cover almost all the central and western parts of the Archipelago. There was probably a heavy vegetational cover over much of the delta, which gave rise to highly

Figure 6.13. The Late Givetian palaeogeography of the Canadian Arctic Islands is shown, together with the outcrops of Lower Devonian formations on Prince of Wales, Cornwallis and Somerset Islands. At this particular time vast quantities of sediment were carried from the Pearya uplands across a coastal plain that was dominated by sand-choked braided streams and a prograding deltaic–marine shelf towards the marine slope. The arrows represent major lines of sediment influx (after Embry and Klovan, 1976).

carbonaceous shales and even to coals. In Late Frasnian time the whole region was uplifted and eroded.

The Peel Sound Formation lies on each side of the northerly directed Boothia Uplift and is comprised of four contiguous facies—a coarse conglomerate facies that is indicative of alluvial fans and interfan areas, a sandstone–conglomerate succession with upward-fining cycles of fluvial origin, a sandstone–carbonate facies of marginal or transitional origin that includes both marine and non-marine strata, and a succession in which the proportions of the different lithologies vary with age. Vertebrate remains are phenomenally abundant at certain levels in the sandstone–carbonate and sandstone–conglomerate facies but persist as scattered items in the marine faunas. There is only scant evidence of a local flora. The Peel Sound Formation arose from the vigorous subaerial denudation of the Lower Palaeozoic and Precambrian rocks of the Boothia Arch, the uplift of the arch being rapid during earliest Devonian or Late Pridoli time. Miall (1973) has suggested that the boulder conglomerates represent alluvial fans that were built up directly in the sea, locally being reworked into beach deposits.

Western Europe

On mainland Europe Old Red Sandstone beds occur as thin and irregular sheets beneath the marine Devonian; commonly they intercede between the marine facies and the pre-Devonian basement. Their narrow outcrops in northern France, Belgium and western Germany have been relatively neglected until recently, but their vertebrate faunas, plant fossils and sedimentary characteristics are similar to those in the British Lower Old Red Sandstone. The uplift of St George's Land during the Caledonian orogeny was followed by rapid erosion and the spread of molasse-like clastics southwards and eastwards in an alluvial regressive phase before the Devonian transgression.

Between the Carpathian Mountains and the Precambrian Ukrainian Shield lies the Lvov Basin in which Devonian deposits reach more than 2200 m. The Lower Devonian continental Drestrov Series rests conformably on the Silurian Tiver Series which contains the famous vertebrate and plant-yielding Chortkov Horizon. The sediments, 1000 m, are alluvial and deltaic with a source area indicated to the south-west.

The Lower Devonian in the nearby Carpathian Mountains and in the Lysa Gora area of the Polish Holy Cross Mountains similarly consists of about 1000 m of Old Red Sandstone facies with vertebrates and abundant plant remains. Here also there is an unbroken sequence upwards from Silurian marine formations in which the progressive swing to continental facies takes place. Following this terrestrial phase, marine conditions became re-established by Eifelian time; an arid climate seems to have prevailed throughout this interval.

The Russian Platform

Devonian rocks cover an enormous area of the U.S.S.R. west of the Ural Mountains (Aronova *et al.*, 1968; Nalivkin, 1973a, 1973b; Sokolov, 1978). The continental facies occur principally in the limited Lower Devonian and as a thin transgressive basal facies in the Middle Devonian over a much greater (wider) region. For example, the Dniester Series in Pre-Dniestrovye contains up to 700 m of red beds with a rich vertebrate fauna while in the Pre-Baltic region the similar Stonishkayi Formation, 100 m thick, also has a Lower Devonian vertebrate fauna. Alluvial or lagoonal red beds higher in the sequence are thin and somewhat scattered but yield vertebrates. Volcanics are also present in this facies in the Donbas. During Givetian time the influx of terrigenous sediment from the Baltic Shield increased and a broad deltaic and lagoonal belt reached several hundred kilometres in width between the Shield and the Moscow region. Limestones and evaporites were locally deposited as marine conditions prevailed, but the whole region was one of rapidly changing physiography with islands and shoal areas. Formations vary in thickness in response to gentle warps in the basin and to varying rates of sediment supply. The entire platform, however, was covered by a carbonate-depositing sea in Early Frasnian time. The inundation was short lived in the Baltic area, as in the Late Frasnian wide distribution of red lagoonal sediments was re-established. From this time on a regressive phase occurred in the Baltic area.

A north-eastern extension of this region, the Timan, contains Middle Devonian red beds and coals. The Upper Devonian of northern Timan includes coals that are up to 65 cm thick. Further north again, on the Novaya Zemlya and Severnaya Zemlya Islands and in the Taimyr Peninsula, continental red beds were deposited as a general erosion of upland areas took place. Severnaya Zemlya reveals the deposits of a semi-land-locked basin with extensive continental sediment and vertebrate and plant remains at the base and, following a marine incursion, a further regressive sequence with vertebrates in the Middle Devonian.

The Altai-Sayan region

In the Altai mountain belt south of Krasnoyarsk in Central Asia very extensive Devonian rocks, resting upon a deformed basement and within fault-bounded basins, reach thicknesses of between 3000 m and 10 000 m. Within the eight basins are alternations of thick red continental strata with volcanic rocks and thinner grey terrigenous carbonates of lagoonal or marine origin. Lower, Middle and Upper Devonian series are represented, commonly by different coeval suites in the different basins. While the Lower Devonian is largely a red clastic and volcanic series with plants, endemic vertebrates and other fossils, the Middle and Upper Series present three or four regular 'cycles' of red beds that are succeeded by grey transitional or marine facies. By Middle Devonian time vast alluvial plains occupied the greater part of this region with volcanic activity around the edge. Several 'sea-size lakes' covered areas of previous upland. Planation was vigorous during Middle and Late Devonian time, despite uplift of the inter-basinal blocks. Plutonic rocks were intruded in Early–Middle Devonian and again in the Late Devonian time. Very full investigations of the regions have been carried out by Predtechensky and Krasnov (1967) and others over the last three decades.

Despite its remoteness, the Siberian Platform has also recently been very closely examined by Soviet geologists and extensive volumes of hydrocarbons are known in carbonate units that are close to the red bed facies of the edge of the platform. Hundreds of metres of continental and lagoonal red beds occur on the north-western, southern and eastern platform margins (Krylova *et al.*, 1968). Transgressions during Siegenian, Eifelian, Late Givetian and Late Emsian times extended marine coverage on to inner parts of the platform while evaporites and phosphatic (detrital bone–breccia) rocks were formed. The region possessed a highly endemic heterostracan (amphiaspid) fauna but also has yielded plants and an extensive marine fauna. Earth movements at the platform margins were responsible for many interruptions to deposition.

A belt of terrigenous clastic deposits can be traced through the Middle East into south Central Asia. Outcrops and formations have been described from Turkey, Iran and Afghanistan and the U.S.S.R. In Turkey's Western Tauride Mountains Upper Devonian red beds with fish remains (Janvier and Marcoux, 1977) may represent a westward extension of the Old Red Sandstone environments that were investigated by Janvier (1977) in Iran.

East Asia

Devonian rocks occur in many scattered localities in South-east Asia but continental facies are thin and limited to the Lower Devonian. In north Vietnam the facies is accompanied by calcareous shales and shales with fish and plant remains. The tectonic setting of these deposits is not clear but is related to the South China fold system, a 'Caledonide' orogenic complex.

In China two 'continental–littoral facies types' have been distinguished among the widespread continental assemblages. They are present in South China, south-east China, and the Qilianshan and Lungmenshan-Tsingling regions (Yang *et al.*, 1981). The Qilianshan type occurs in intracontinental basins — red clastics with volcanic beds and many local unconformities. The biota is mainly plants and vertebrates, cosmopolitan and endemic. The Qujing type appears to be more characteristic of external basins, being mainly continental red beds that are intercalated with thick marls and sandstones of pale colours. The Lower Devonian representatives of this type contain the 'oriental' vertebrate fauna as well as cosmopolitan taxa and plants. In Yunnan Province, South China, some 2100 m of Qujing type extends from the base of the system to the Givetian. Sections in Gwangdong and Hunan Provinces reveal thick formations of Middle and Upper Devonian age. South and east China has Lower and Upper Devonian red bed formations with vertebrates, while the Qilianshan type in central China (as at Ning Xia) extends over Middle and Upper Devonian ages. In the Lungmenshan the Lower Devonian Pingyipu Formation (220 m) yields plants and a large and unique vertebrate fauna.

These formations are associated with Caledonide fold systems (South China, Kumchun, Nanshan, Tienshan and Inner Mongolia) as post-orogenic deposits, and as basal beds in transgressive sequences. Lagoonal and transitional–marine environments are well represented in the Qujing type

AGE		SOUTH CHINA REGION				
		Chuifengshan, Chutsing, Yunnan	Liujing, Kwangsi	Tiaomachien, Changsha, Hunan	North Kwangtung	Yudu, South Kiangsi
		C_1	F	C_1	C_1	C_1
UPPER DEVONIAN	D_3^2	Zaige Formation* Grey massive limestone 460 m	Rongxian Formation*	Yuelushan Formation Yellow quartzite and sandstone 80 m	Maozifeng Formation*	Xiashan Formation Sandstone, quartzite 300 m
						Shanmentan Formation* Sandstone, mudstone 200 m
	D_3^1			Shetienchiao Formation*	Tianziling Formation*	Zhongpeng Formation Sandstone, shale 270 m
MIDDLE DEVONIAN	D_2^2	Haikou Formation Yellow sandstone, quartzite and shale 15–200 m	Tungkangling Formation*	Chitzechiao Formation*	Chitzechiao Formation*	Yunshan Formation Quartzite, sandstone 70 m
		Sanshuanghe Formation Yellow sandstone and grey dolomite 200 m		Tiaomachien Formation Red sandy shale and shale, yellow quartzite and conglomerate	Dahepo Formation 500–600 m Red and yellow sandstone and shale	
	D_2^1	Chuandong Formation Yellow sandstone 80–90 m	Najiao Formation*		Guitou Formation 500 m Yellow, red quartzite and sandstone	
LOWER DEVONIAN	D_1^3	Xuhiachong Member Red and yellow sandstone and shale 870 m				
	D_1^2	Guijiatum Member Red sandstone and shale 300–360 m	Yukiang Formation* Nakaoling Formation*			
	D_1^1	Xitun Member Grey argillaceous beds 300 m / Xishancun Member Yellow sandstone and green shale 320 m / Miandiancun Formation Black shale 30–70 m	Liukankou Member Red sandstone 150 m / Hengxian Member Purplish-red mudstone with dolomitic limestone 110 m / Lingli Member Sandstone and conglomerate 70–100 m			
		Upper Silurian Yulongsso Formation	Є	S	O	

(vertical labels: Chuifengshan Formation in Chuifengshan column; Lianhuashan Formation in Liujing column)

Figure 6.14. A correlation table of the Devonian continental sedimentary formations in China illustrates the palaeoenvironments is indicated and the fossils include macroplants, palynomorphs invertebrates and both (after Yang *et al.*, 1981).

but detailed sedimentological studies and basin analyses have only recently begun. Huang (1978) regards the Devonian molasse of the Nanshan as the most typical post-orogenic deposits, with volcanic rocks contributing heavily to the pile. Figure 6.14 shows the Old Red Sandstones of China.

Japan has no development of non-marine Devonian but the characteristic Late Devonian plant *Leptophloeum* is known from two localities in association with marine fossils.

Australia

Devonian continental deposits occur in the intracratonic basins of central and Western Aus-

SOUTH-EAST CHINA REGION		CHILIENSHAN REGION	AGE	
South Kiangsu, West Chekiang, South Anhui	Xiushui, North Kiangsi	Niushoushan, Zhongning, Ningxia		
C_1	P_1	C_2		

Wutung Group   50–180 m
  Yellow quartzite and
  grey shale
	Wutung Group 15–50 m
	  Yellow quartzite
	  and grey shale
		Zhongning Formation
		  Upper part: purple and red
		  sandstone and shale
		  Lower part: red and purple conglomerate,
		  quartzite and sandy shale 26–50 m
			D_3^2
			
			D_3^1
				UPPER DEVONIAN

		~~ Unconformity ~~~~~~~
		(Ningxia Movement)
		Shixiagou Formation
		  Upper part: purple sandy shale and sandstone
		                180–200 m
		  Lower part: yellow sandy shale, quartzite and
		  conglomerate 70–80 m
			D_2^2
			
			D_2^1
				MIDDLE DEVONIAN

Maoshan Group
  Red, purple and green
  sandstone, shale and
  quartzite   0–500 m
	Xikeng Formation
	  Upper part: green and purple
	  sandstone and shale, with
	  megafloras and agnatha
	  fragments 120 m
	  Lower part: purple, red,
	  green sandstone and shale
	           180 m
			D_1^3
			D_1^2
			
			D_1^1
				LOWER DEVONIAN

S_3	S_3	ϵ_2		

profusion of units now known in widespread parts of this enormous country. A very wide range of endemic and cosmopolitan vertebrates. Names with asterisks are formations used as stage stratotypes

tralia and also on the margins of the Tasman Geosyncline in the eastern third of the continent (see figure 6.15). In the Canning Basin marine clastics, carbonates and evaporites were deposited as the Middle Devonian Carribuddy Formation which perhaps is largely of pre-Devonian age, but the succeeding Tangaloo Red Beds, 750 m thick, is a continental facies with possible aeolian components. It is overlain by Middle Devonian evaporites and carbonates. At the northern end of the basin Late Devonian red beds with plants and fish remains are present. Aeolian and lacustrine beds occur in the Amadeus Basin of central Australia with spores and vertebrate remains confirming a

Late Devonian age. In the Duline and Toko synclines to the north-east further red beds yield plants and vertebrates of Lower and Upper Devonian ages. Alluvial and lacustrine sedimentation appears to be responsible for all of these formations in a system that drains the present central Australia area to the south-east.

Occurrences of continental facies adjacent to or within the Tasman Geosynclinal Belt are very numerous. Many contain plant remains and vertebrate fossils, especially of Middle and Late Devonian ages. While many red bed clastic formations are thin others are in excess of 2500 m and may be associated with volcanic rocks. Environments that

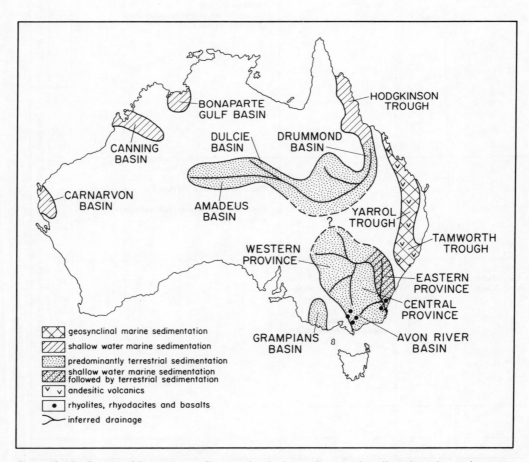

Figure 6.15. Basins of Devonian sedimentation in Australia are primarily epicontinental except on the eastern margin of the continent. In all of them some terrestrial and non-marine deposits were laid down and with these are associated abundant and diverse vertebrate faunas. See also figure 7.15 for the general palaeogeography of Australia during the period.

range from fluvial through estuarine and lacustrine to lagoonal and marine are indicated. Marine conditions persisted in the east throughout the period but unrest in the geosynclines was almost continuous, especially during the mid-Devonian Tabberabberan Orogeny. Upper Devonian non-marine rocks are especially widely distributed in south-eastern Australia. Tentative models of braided river and shoreline palaeogeographies have been suggested for these formations and thick accumulations of volcanic rocks are noted.

Antarctica

Within the great clastic sequence known as the Beacon Supergroup of eastern Antarctica and the Transantarctic Mountains the Devonian component is dominated by red beds with abundant plant and vertebrate fossils. Lower Devonian vertebrates have been recognised in the Horlick Mountains and marine invertebrates have also been found there; elsewhere they are of Middle and Upper Devonian age. The coarse red beds throughout this huge area seem to represent vigorous infilling of elongate depressions that developed along the Ross orogenic belt. They were deposited in innumerable upward-fining cycles and with the expansion of the area of clastic deposition much sediment came in from the shield to the west, while a marine seaway crossed the continent with constantly shifting shorelines. In the (Upper Devonian) Aztec Siltstone of southern Victoria Land the alluvial sequence contains palaeosols that are similar to those in the red beds of the North Atlantic area and to modern pedocals (McPherson, 1979). Devonian rocks are now also known in Marie Byrd Land, and volcanic activity occurred in the Ross Sea region.

South America

The Lower Devonian of the Falkland Islands includes a continental fluvial facies which appears to be close to the sediment source and the Devonian also bears a strong resemblance to the Cape system of South Africa. Higher up plant remains in a 100 m micaceous flagstone series suggest a Middle Devonian age.

Elsewhere in South America red-coloured clastic rocks contain scarce plant remains but so far no vertebrates have been recorded. Sedimentological studies of the Devonian in South America have not yet confirmed true continental deposits that are comparable to the Old Red Sandstone elsewhere.

These continental facies record processes and events no less effectively than do the marine. Devonian continental suites arose in response to uplift, orogeny and the development of sedimentary basins within and adjacent to orogenic belts and, ephemerally, in the van of marine transgressions. They include indicators of both humid and arid climatic conditions, periods of palaeosol formation, fluvial and mass-movement of regolith on locally impressive scales, continuing orogenic uplift of source areas and, withal, a general continuity of continental and intertidal environment in Middle and Late Devonian times that allowed rapid and extensive vertebrate migration.

7

Paralic and Cratonic Facies

Under this heading are grouped strata that, in area, constitute the greatest part of Devonian outcrops today. Ronov *et al.* (1980), however, note that their volume is considerably less than that of the 'geosynclinal' deposits. The greater spread of the shallow-water, platformal or shelf sediments is evidence of the Devonian transgressions indicated in figure 1.4, and the greatest wealth of data on shallow-water (marine) environments is contained in them. Despite their accessibility, however, the shallower marine environments remain far from fully understood and their abundant biotas leave a *relatively* small palaeontological record.

During Devonian time transgressing seas linked active cratonic basins with those at the continental margins, extending oceanic-water circulation and the spread of terrigenous, carbonate and even evaporite deposition. When regression occurred prograding sediment was laid down or erosion removed the recently deposited material. In each case the changes were brought about within a migratory zone that is subject to very energetic processes: wave, wind, tidal current, climatic and biological activity is unceasing and variable. They occur in a transitional realm between the continental and the open-water marine environments, and embrace deltas, shorelines and lagoons, which are highly sensitive to changes of sea level and to earth movement. A well-documented Middle Devonian example of the vagaries of sedimentation during a brief transgressional phase within a regional regression in New York State (McCave, 1973) is said to be related to a total sea level rise of no more than 18 m. The environments involved range from tidal flat to open shelf and slope.

The term 'shelf', often applied to such areas of deposition, includes both continental shelves and partly enclosed seas such as the North Sea or Baltic Sea today. However, the enormous spreads of Palaeozoic shallow-water sediments are of a quite different order of magnitude from the areas of present continental shelf. Nothing today compares with the Palaeozoic flooding of the great cratonic interior of North America or parts of North Africa. Thus to explain the depositional regimes responsible for the 'layer cake' accumulations of shallow-water sediments, between uplands and the edge of the continental shelf (*paralic* or *epicontinental*, see Heckel, 1972) and those spread so evenly across the continental interiors (*cratonic*) (see figure 7.1) there is a place for models that are based upon geological rather than actualistic concepts. A major hindrance to the interpretation of suites of ancient shelf deposits is the lack of a satisfactory way of determining the depth at which the sediments were laid down. Fossils may be of some help but it seems to be relatively little. It has to be admitted that we are still some way from understanding the hydrological mechanisms that are responsible for distributing Recent sediments across the continental shelves and some long-held ideas on the subject are proving inadequate. Nor must it be assumed that either the cratonic or the paralic seafloor itself is necessarily tectonically immobile over even short periods of time.

These categories of depositional environment both come within the meaning of the term 'epeiric' (see especially Hallam, 1981). Within both we may expect a similar sequence of environments to be present though not all members of the sequence may occur. The many lithological characteristics of strata that are formed within these depositional environments are generally well known, but puzzling features occur in many ancient deposits.

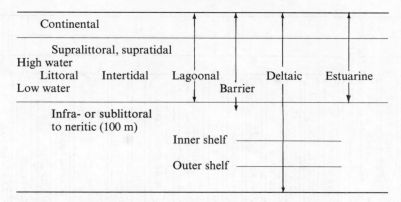

Figure 7.1. Depositional environments within the paralic or cratonic seas.

Tidal flat and sabkha environments, coastal mudflats and aeolian sandbanks may characterise the topographically highest areas, while spreads of sheet sands and muds may indicate the lowest. Sediment supply is an important factor in delimiting these sedimentary environments: storms, and other climatic and hydrological phenomena, may interrupt the otherwise 'normal' processes of sedimentation. The expenditure of energy ranges from very low and even in the outer shelf to very high and fluctuating in the intertidal reaches.

The paralic (neritic) facies belts border the continents, some being narrow and descending to the continental edge relatively steeply; others are broader and less inclined. The cratonic facies assemblage may have a much less regular aereal arrangement, generally with extremely low bottom gradients for the most part. Both may be profoundly affected by synsedimentary tectonic movements (see for example Krebs and Wachendorf, 1974; Solle, 1976), which may involve faults and/or flexures.

Goldring and Langenstrassen (1979) make the point that Devonian open shelf and near-shore clastic facies possess no distinctive attribute apart from their biota. Exactly similar sedimentary facies may be found in the Silurian or the Carboniferous. The fossils themselves (see figure 7.2) may, nevertheless, greatly aid our interpretation of the marine environment (Heckel, 1972).

Shelf Morphology

Ancient marine shelves (no less than modern ones) were relatively mobile in the sense that the morphology was undergoing continuous change. Changes result from different rates of subsidence from place to place, differing supplies of sediment, wave and tidal regimes, eustatic change and shoreline progradation and the flux and redisposition of sediment. The ideal model is one with a gentle slope seaward without sharp breaks in gradient and with the highest energy confined to a relatively narrow zone near the shore. Normal wave base, however, may coincide with an offshore break in slope and set up an offshore zone of high energy. Tides and currents have been found to exert an unexpectedly strong influence in open offshore parts of the shelf. Local biological activity may affect the generation, deposition or erosion of sediment, especially where carbonates are involved (see page 137).

Both from stratigraphic studies on land and on the present continental shelves it is clear that rates of accumulation may differ across the width of a shelf with isopachs marking troughs that are parallel to the coast or the edge of the shelf. On the seaward side of these may occur reefs or other carbonate buildups. With the passage of time the troughs and the edge of the shelf may migrate, moving either towards the strand-line or towards the continental slope. Fault activity in the basement beneath shelf bars or reefs has been held to be repeatedly responsible for axes of uplifts and 'arches', so influencing sedimentation profoundly and affecting the distribution of the biota. During strong transgressions the effects of such positive areas upon sedimentation may be short lived; with regression they may serve to separate a once-

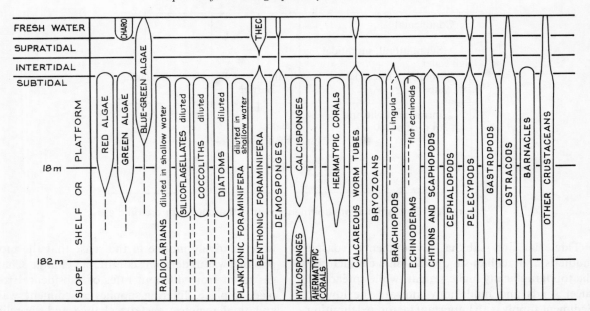

Figure 7.2. The vertical distribution of Palaeozoic marine invertebrates in epeiric seas may have been comparable to that of modern groups in Recent seas. Their grouping into communities seems to have been primarily dependent upon depth, nature of substrate, temperature and salinity but the intertidal zone may contain much skeletal debris (after Heckel, 1972).

continuous environment into parts that then have different sedimentary histories. It has been suggested that, with strong regression, the inner shelf may experience rapid progradation while the outer shelf is little affected.

Far from being featureless, the cratonic regions reveal that deposition was influenced by arches and troughs, domes and basins of varying but possibly very sharp relief, which were persistent for many millions of years even where far removed from the mobile crustal zones at the margins of a continent. The mid-Palaeozoic configuration of the cratonic surface of North America provides good examples where specifically Devonian positive and negative areas ('highs', 'arches' and 'basins') have been identified west of the Canadian Shield (see figure 7.3). Faults play some part in defining these structures but the origins of the stresses released by fault movement or flexuring seem to lie in the continental crust below. Walcott (1970) postulated isostatic relief to account for the continuing activity of the Boothia Arch and this mechanism has been involved elsewhere. Krebs and Wachendorf (1974) and Krebs (1977) have emphasised that vertical

tectonic components were highly influential in the evolution of the epeiric depositional basins of Europe during Palaeozoic times.

Paralic Shelf Facies

Clastic shelf sediments range from pebble-beds to fine muds; carbonates may be of many kinds (see chapter 8), sorting may be from good to poor and maturity may be of similar range. The distribution of these materials may be apparently regular and in belts parallel to the coastline, or not. Nevertheless, despite the minor vagaries of distribution, many shelf assemblages (see figure 7.4) do conform to a broad spatial arrangement suggested by Langenstrassen (Goldring and Langenstrassen, 1979, p. 84).

For Devonian examples it has been observed that the *inner* part of a shelf is dominated by mixed barren and shelly sandy mudstone facies that contain locally a high proportion of red and green sediments, with the wide range of grain sizes and the relatively poor sorting that is associated with immaturity. Bioclastic limestones are not common among

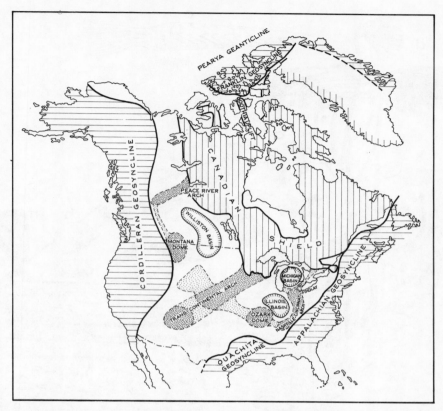

Figure 7.3. The character of the sheets of Palaeozoic sediment spread over a craton depends largely upon the way in which the craton surface subsides or rises. The structures of the relatively stable interior of North America in mid-Palaeozoic time are shown. Darker-shaded areas of the arches were more persistently rising than the others, where movement occurred only now and then (after Clark and Stearn, 1960). The effects were especially felt in the deposition of carbonates, evaporites and clastic rocks during the Devonian.

these sediments. Among sedimentary structures that are present are large-scale cross-stratification, channel-forms, intraformational and extraformational conglomerates, oscillation and current ripples, and desiccation features. Large ball and pillow structures may occur. The fauna represented may include fish and eurypterids, bivalves, ostracods and a few brachiopods (bedding surfaces may reveal only one species), and locally abundant plant remains. Trace fossils are conspicuous and include predominantly vertical structures such as *Diplocraterion*, *Lennea*, *Skolithus*-like forms, escape structures and, in muddy sediment, *Spirophyton* (see Goldring, 1971).

The *outer* shelf shows an essentially bioturbated and shelly mudstone and siltstone facies which lacks red and green sediments but which displays a small range of grain size and has relatively good sorting. Interbedded calcareous units include sandstones, coquinites and thin limestones. Individual beds are remarkably persistent, shales and sandstone alternating. Sharp soles and graded tops to beds are common and the internal structure may include small-scale cross-lamination and convolute bedding. A high diversity of mainly stenohaline groups makes up the biota, and bioturbation may be intense. There is normally a gradual transition from inner to outer shelf facies, depending upon types of sediment, rates of supply and current activity (see figure 7.5).

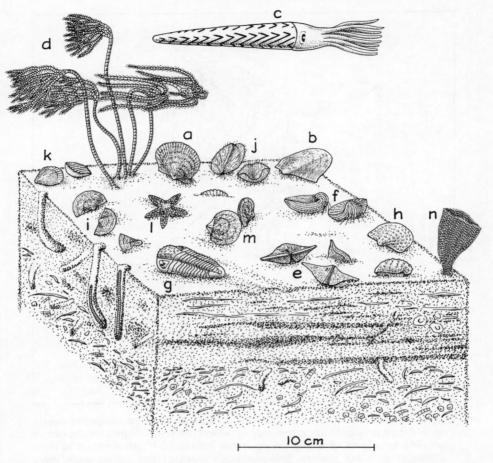

Figure 7.4. An Upper Devonian offshore shelf community, a diverse fauna including (a) *Pterinopecten*, (b) *Ptychopteria*, (c) *Actinoceras*, (d) crinoids, (e) Cyrtospirifer, (f) *Mesoplica*, (g) *Phacops*, (h) *Productella*, (i) *Chonetes*, (j) *Athyris*, (k) *Camarotoechia*, (l) *Palaeaster*, (m) *Schellwienella* and (n) *Fenestella*, on a sandy bottom (after Goldring, in McKerrow, 1978).

Beach and *shoreline* deposits have now been identified in many Devonian formations. Sutton *et al.* (1970) described a sequence of facies from near-shore through delta platform (inner shelf) to delta front sands and outer shelf in the Upper Devonian Sonyea group of New York State and a muddy shoreline sequence from Pennsylvania has been postulated by Walker, R. G. and Harms (1971). No coastal barrier sands or lagoonal silts and muds can be identified positively here. It is thought that the shelf was very extensive and shallow, comparable to a situation described by McCave (1968) also in New York State. Organic productivity in the beach

environment is not high and most fossils present in beach sediments are either infaunal types or exotic.

Restricted bay and lagoonal facies of both clastic and carbonate lithologies are numerous in Devonian stratigraphy. The critical factor was salinity which influenced the facies biota. Greenish mudstones and siltstones and fine-grained sandstones are common lithologies as are shelly or crinoidal limestones and siltstones. Some sheet sandstone may be present and wave ripple-marks are common. The faunas of brachiopods, bivalves, ostracods and bellerophontid gastropods are of low diversity: trace fossils may be common. Where evaporation

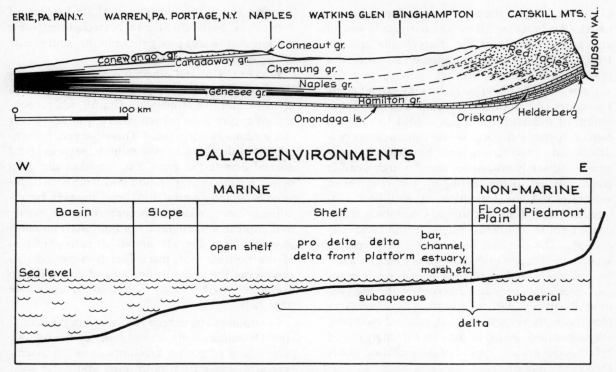

Figure 7.5. Facies changes in the Devonian of New York State, the classical section as offered by Schuchert and others in 1924. Black represents black shale units, unshaded is calcareous shale, grey and red shales and fine sandstones; the red facies ('Catskill' and 'Pocono') is typically medium-grained to coarse-grained sandstones and conglomerates. The vertical thickness is much exaggerated. Below is an interpretation of the palaeoenvironments represented (after Rickard).

rates exceed influex, salt concentration rises and evaporites may be deposited, especially if terrigenous sediment input is low and the land nearby is arid.

Tidal and intertidal deposits (see Ginsburg, 1975; Ovenshine, 1975) may be preserved where transgression, or regression with a copious supply of sediment, or an efficient sediment-trapping mechanism operates. Laminated sediments and many sedimentary structures are characteristic and marine fossils dominate (Howard, 1972); bioclastic debris may be abundant. Tidal-flat environments are ephemeral in that they result from a delicate balance between erosion and deposition and the vertical tidal range. Several Devonian tidal deposits appear to have formed under weak tidal conditions; others to have accumulated where tidal or disruptive storm influence was strong and periodic. Laporte (1967a, 1967b, 1969, 1975, 1979) was

unable to determine the cause of water-level fluctuations that controlled (Lower Devonian) Manlius carbonate tidal-flat deposition—astronomic tides, wind tides, storms, monsoon climate—nor was he able to estimate the length of the intervals between floodings. The modern stromatolite tidal deposits of Shark Bay, Western Australia, provide a model for ancient stromatolites, and cryptalgal and fenestral carbonate fabrics are taken by many authors as reliable indicators of an intertidal origin. Mountjoy (1975) regarded fenestral laminae as stirred-up sediment that is spread by abnormally high tides over lagoonal (tidal) flats, and concluded that the Upper Devonian thick laminated facies in Alberta could not possibly represent a true tidal cycle.

The faunas of Devonian intertidal environments are not well known but probably they conformed to the general pattern that was typified by the Silurian *Lingula* community described by Ziegler, A. M. *et al.* (1968). *Lingula* itself is common locally: other inarticulate brachiopods such as *Barroisella* were of an uncertain mode of life. Eurypterids and ostracods may have been widespread and successful in these environments, bivalves certainly were.

Open shelf and shoreface facies may extend many kilometres out from the shoreline and the distribution of different lithologies is held to reflect a meteorological influence. Where calm weather predominated, fossiliferous and bioturbated mudstones and sandy mudstones result; rough weather conditions led to the stirring-up and removal of substrate and the formation of shell-beds (coquinities) and sheet sandstones. Commonly these coarser-grade sediments dominate the local successions. The contrast between the two types of sediment is sharpest in shallow waters; deeper-water environments are less affected. Many fossils in the shallower areas may be preserved in the life position. Bivalves in particular seem to have been very plentiful both as epifaunal and infaunal members. Rhynchonellid, spiriferid and other strongly ribbed brachiopods are also typical of the shallower sandy Devonian seafloor.

Disturbance of the water during rough weather induces aeration and the supply of nutrients to deeper environments, to the benefit of the biota; it may also introduce fresh sediment and hence a new potentially productive substrate may be formed. Studies of the palaeoecology and faunal community structure in such facies are relatively few so far. Goldring and Langenstrassen (1979) and McGhee and Sutton (1981) have noted some of the more significant.

Devonian Shelf Communities

Pioneer work on Devonian benthonic shelf (brachiopod and other) communities in eastern North America has revealed relationships to sedimentary facies that are not greatly dissimilar from those of the Silurian. The influence of a muddy prograding delta front rather than depth of water alone was a prime factor that influenced the composition of the several bottom communities. This work by Sutton *et al.* (1970), Bowen *et al.* (1974) and Thayer (1974) is summarised in figure 7.6. McGhee and Sutton (1981) noted that benthonic marine

communities in the Java Group (Late Emsian) of New York State could also be recognised some 500 km southwards along strike in the central part of the Appalachian region.

Boucot (1975) and others have studied the benthonic communities over much of the shallow-sea areas within the American Craton where conditions appear to have been relatively uniform over extraordinarily wide areas. There the epicontinental seas evidently retained a salinity that was about normal despite the great distance from the open ocean. Excessive evaporation may have been countered by abundant rainfall. Where, however, hypersaline bodies of water did occur an effective barrier to the spread and migration of bottom communities would arise. Periodic anoxia or eutrophication of the bottom took place for short periods, so wiping out the previous faunas and giving rise to distinctive marker beds in the stratigraphy (see page 121).

Communities have been recognised in the Rhineland Devonian, again on the basis of the relative proportions of certain brachiopods in the assemblages of fossils from both inner and outer shelf regions. Groups that were more tolerant of high levels of turbidity and sedimentation characterise the inner shelf while a more normal marine fauna was present in the outer shelf. Trace fossils are abundant only very locally and the range of types is small. The outermost part of the shelf favoured the preservation of ichnofossils to a greater extent than the more turbulent centre and inner parts.

Faunal communities in carbonate shelf facies appear to be more diverse than those of the clastic shelves but detailed work on these highly productive environments is fraught with the difficulties of interpreting the palaeoecological relationships. Diagenetic changes add to the problems of recognising the original nature of the substrate.

Cratonic (Epeiric) Deposition

Characteristic of the Devonian rocks that are spread over much of the interior of North America, Asia and elsewhere is the relatively low ratio of clastic sediments to carbonates. These epicontinental seas, situated throughout much of the tropical zone, were shallow and had remarkably low depositional slopes, perhaps only 0.2 m per km or less.

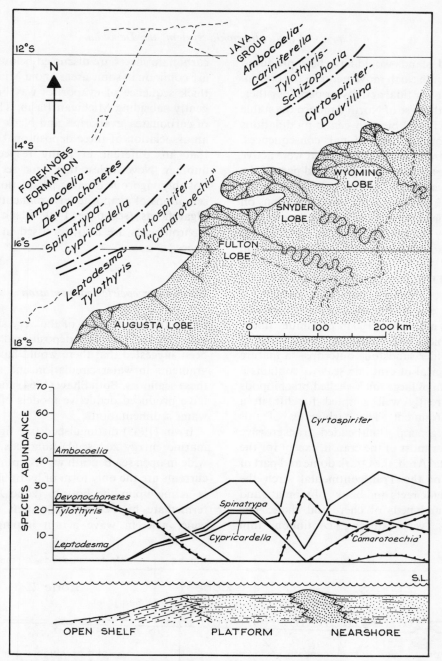

Figure 7.6. Late Frasnian environments in the New York–central Appalachians and their brachiopod-dominated benthonic communities have been reconstructed by McGhee and Sutton (1981). The Foreknobs Formation onshore–offshore succession of marine environments and species distributions is shown below the map. The *Ambocoelia-Devonochonetes* Community (largely epifaunal, unattached and immobile brachiopods) lived under conditions of low energy and a fine-grained substrate. The *Spinatrypa-Cypricardella* Community inhabited the broad delta platform above the open shelf and is an epifauna with attached and unattached immobile forms. The *Cyrtospirifer-Camarotoechia* Community is largely epifaunal but infaunal mobile forms are present. Where not mobile, taxa are firmly attached and with robust shells for life under current-dominated conditions. The *Leptodesma-Tylothyris* Community, in the south only, occupied lagoonal habitats and included attached pediculate or byssate species, most of which were probably euryhaline.

Shoal areas and islands would have been produced by even very gentle earth movements in the craton: by strong current or tidal activity or perhaps other, rare, events. In the case of North America the stable interior was in fact subject to gentle modulations that had important depositional consequences. Figure 7.3 illustrates the structures that were active at some time or another during the mid part of the Palaeozoic. Much of the evidence for these structures is from the thinning of various formations on the crests and flanks of the uplifts. The basins may exhibit not only thicker stratigraphic columns but also different facies, as in the case of the Williston Basin.

The Lower Devonian Oriskany Sandstone is a good example of 'blanket' sandstone, covering a very wide area east of the Mississippi Valley and reaching to no more than 100 m maximum thickness. It is a typical orthoquartzite, the sand being derived from older sandstone outcrops—a mature sand that is typical of cratonic seas. It contains a distinctive fauna of large thick-shelled brachiopods that are apparently well adapted for life in a turbulent environment. The carbonates of this region are in general thin-bedded transgressive sheets that cover most of the craton, except for the Transcontinental Arch, the Ozark dome and part of Kansas. East of the Transcontinental Arch the carbonates include reefs and banks of tabulata and rugose corals and beds of chert and shale. The origins of these very extensive, continuous thin carbonate sheets are discussed below. Black shales are common in some areas and a Middle Devonian thick sequence of evaporites was built up in the gently subsiding Michigan Basin. The association of carbonates, evaporites and black shales implies an association of three depositional environments that are different in many respects from one another. Nevertheless they each seem to indicate possibly vigorous organic production. Recent observations of evaporite environments disclose rich algal floras (see page 149). Black shale anoxic environments are also discussed in the next few pages.

Clear-water epeiric sedimentation

Given the great width of the very shallow epeiric seas and the very low depositional slopes, it has been suggested that there would have to be sharp gradients in water circulation and agitation over these seafloors. Both Shaw (1964) and Irwin (1965) have produced deductive models of epeiric clear-water sedimentation.

Irwin (1965) distinguished (see figure 7.7) three marine energy zones: (1) hundreds of kilometres wide in open sea beneath wave depth where marine currents are the only form of hydraulic energy that is acting upon the bottom (zone X); (2) an intermediate high-energy belt, tens of kilometres wide, wherein wave action impinges upon the

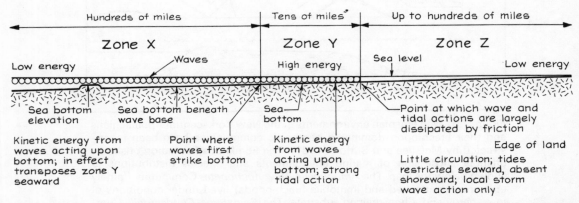

Figure 7.7. Irwin's (1965) model of energy zones in epicontinental seas. While suitable to explain the great spreads of siliciclastic and evaporitic materials (zone Z) over the Palaeozoic cratons, there appears to be no analogue of this zone in the present-day world. Carbonates and organic reefs in zone X (and Y) are also very widespread. In zones X and Z the depositional interface may have been so undisturbed as to promote stagnant conditions, with black shales resulting (see figure 2.6).

seafloor and extends into the limit of tidal action (zone Y); (3) an extremely shallow, low-energy zone, tens to hundreds of kilometres wide, where there is essentially no tidal action and little circulation of water, the only wave action being produced by local storms (zone Z). Kazmierczak (1975) postulated a similar environment for the deposition of Late Devonian fine-grained bituminous black limestones in Poland.

Leaving aside any terrigenous clastic influx, the dominant sediments in zone Y are basically biogenic in origin, sand sized or coarser; in zone Z chemical sediments originate; in zone X sediments are mainly fine-grained detritus from zone Y.

During a marine transgression these three zones move progressively shorewards, each zone overlapping older sediments as it migrates. Eventually a vertical sequence of sediments is formed which mirrors the horizontal arrangement of sedimentation zones. During regression the offlap replaces overlap. As these movements take place the age of the deposited material becomes progressively younger in the direction of migration. The resulting geological formations are diachronous.

Laporte (1967b) examined the Lower Devonian Helderberg Group of New York State as an example of a transgressive carbonate sequence that originated possibly as Shaw's and Irwin's models would suggest. The diachronous nature of the individual stratigraphic units within the group was known. The Helderberg Group has an extensive outcrop and reaches a maximum thickness of about 110 m in eastern New York State. Laporte's work resulted in an interpretation of the depositional environments of the four formations of the Helderberg Group, as shown. The Manlius (see figure 7.8), with mud-cracked dolomitic and barren laminated deposits, may have originated in tidal-flat conditions that are similar to those described for the Bahamas and South Florida. In the Coeymans true marine neritic conditions operated and in the Kalkberg and New Scotland formations the evidence is of sediment that was deposited below the normal limit of wave and current reworking.

The unbroken sequence from tidal-flat lagoon to quiet marine waters was demonstrated. Reviewed in terms of Shaw's and Irwin's models, the Helderberg sea provided a set of zones in an east–west direction (that is, parallel to the postulated coastline). The

Kalkberg and New Scotland formations were deposited below wave base (that is, in water about 12–15 m deep). Within the 110 km wide facies belt of the Helderberg this indicates a gradient of less than 0.3 m per km. To the east open deeper water prevailed. Land-derived sediment from a source further east entered this zone. In the west lay the Manlius complex of tidal flats and shoals. Four similar contemporaneous environmental regimes are suggested for the Helderberg outcrops in other Appalachian states. The transgression was relatively slow with regional subsidence that permitted the accumulation and migration of the depositional environments described. Laporte's work seems to uphold the model for clear-water epeiric sedimentation.

Black Shales

As mentioned above, black shales may form conspicuous units within epeiric shelf sequences, where in general they tend to be thinner and distinct from those of deeper-water environments. Epeiric black shales may be carbonaceous where plant remains are abundant, or bituminous where marine productivity was very high. Rarely is more than 7 per cent of the rock composed of carbon; kerogenous matter may not reach as much as 14 per cent. As in the Appalachian regions, black-shale units may extend from basinal areas into cratonic shelf or deltaic suites, indicating rapid marine transgression and anoxic sea-floor conditions. While they are good stratigraphic marker bands, their great geographical spread does require explanation (see figure 7.9).

Within the shales diagenetic carbonate, pyrite or chert nodules occur; trace fossils are very rare or absent and body fossils are restricted in type and distribution. Pelagic species dominate. In some instances the taxonomic range is great but commonly it is restricted to a few taxa, one or two being locally very abundant. Bedding planes may be covered by a single species. Some Devonian black shales contain rich conodont faunas and huge numbers of individual conodont elements.

Depth of deposition does not seem to have been a prime factor. The essentials are the physically undisturbed condition of anoxic water, a reduced

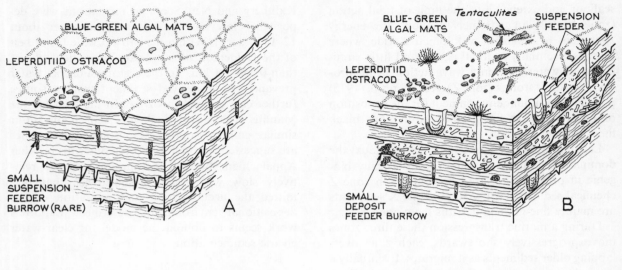

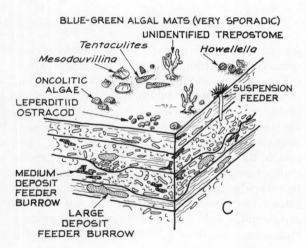

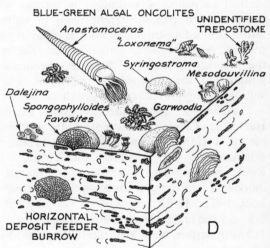

Figure 7.8. Communities in the Manlius Formation of New York State (after Walker, K. R. and Laporte, 1970). (A) A supratidal environment with heavy algal colonisation and restricted numbers of invertebrates. (B) High intertidal community composition is more varied but still low in taxonomic diversity. (C) At the low intertidal level a wider variety of epifaunal benthos is apparent. (D) Subtidal communities may reach a state of high taxonomic diversity, largely amongst the epifaunal elements.

supply of clastic debris, and high organic productivity in the oxygenated zone above. High organic productivity in the surface layers leads to a rain of material which, decomposing on the undisturbed substrate, depletes the dissolved oxygen there. Where low tidal ranges and poor circulation of water at depth obtain, stagnation of the bottom may be a relatively permanent rather than a seasonal feature. Well-laminated black shales can indicate seasonal changes in debris supply and decomposition. In the deeper seas and open oceans anaerobic conditions may have prevailed for the same reasons. Indeed, the great extent and thickness of many Lower Palaeozoic black shales suggests that they are oceanic rather than cratonic or marginal to continents. Berry and Wilde (1978)

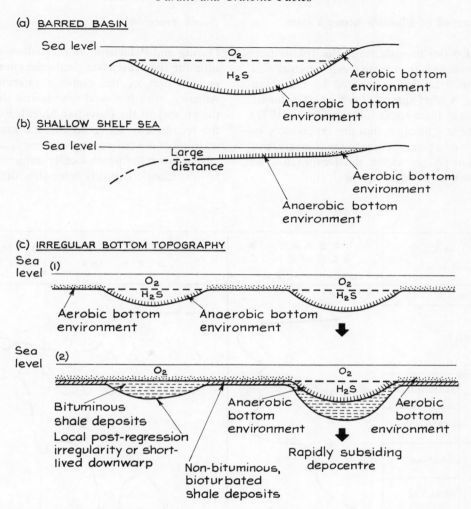

Figure 7.9. Black shales, more specifically bituminous shales, are widespread in the Devonian epicontinental and basinal successions. They are usually held to lie close to the base of a transgressive sequence, either on a small scale as in New York State or on a much larger one (Chatanooga Shale and others) as throughout the Appalachians and Mid West. Several models have been proposed for the depositional environments of black shales (Hallam and Bradshaw, 1979) all of which involve the creation of bodies of anaerobic still water below the aerated layers.

suggest that the unventilated condition of the ocean floors was ameliorated by the slow flow of glacially cooled oxygen-rich oceanic waters from high latitudes equatorwards during the Ordovician and Silurian glaciations. By Devonian times there may have been a halt to this process as glaciation diminished. Flow from cooler regions into the deeper parts of the Devonian latitudinal seas of the tropics may have been restricted by the changes in continental positions and the great productivity of their surface layers would tend to reduce levels of oxygenation over deeper seafloors. The temporary spread of anoxic bottom conditions on to the cratons and continental margins, although it may have little to do with oceanic conditions, remains puzzling. Today these shales are an important source of oil and gas in the Appalachian area.

The Spread of Shallow-water Facies

An aspect of major importance in the distribution of Devonian rocks is their transgressive nature and the enormous area that is covered by Devonian shallow seas. A brief survey of some of the larger outcrop areas of these rocks follows. House (1977) has outlined the difficulties that are experienced in attempting to give precise dates to the transgressive and regressive phases within the Devonian. The general pattern is, however, quite clear.

North America

During mid-Palaeozoic times shallow seas responsible for the Kaskaskia Sequence spread over the greater part of the cratonic interior of North America, with low land masses and shoal areas in the vicinity of the Canadian Shield. Even far from the tectonically active continental margins gentle warping of the basement plus some fault movement gave a low relief which locally influenced sedimentation. Clastic deposits were thin orthoquartzitic

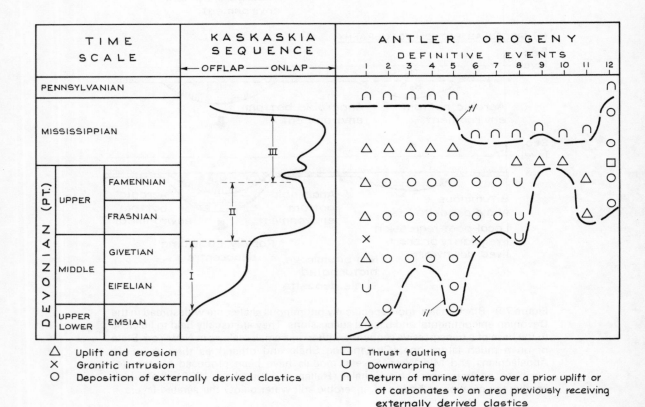

Figure 7.10. The Kaskaskia Sequence in western North America shows the influence of the Antler Orogeny in Johnson's (1971) figure. He uses the term depophase to indicate a time span of epicontinental sea expansion or recession in response to epeirogeny and eustasy. The regions in which the Antler orogenic effects are plotted are: 1. Northern Axel Heiberg and Ellesmere Islands; 2. Central Ellesmere Island; 3. South-western Ellesmere Island; 4. Eastern Melville Island; 5. Western Melville Island; 6. Brooks Range, Alaska, Romanzof Mountains, and northern Yukon territory; 7. Mackenzie Valley, North-west Territory; 8. McDame area, northern British Columbia; 9. Southern British Columbia; 10. South-eastern Alaska; 11. Klamath Mountains, northern California; 12. Nevada–Idaho. The lines *il* and *tl* are somewhat irregular because they go to and fro across the geosynclinal belts and because there is a southern migration of inceptive events through time. In eastern North America the Acadian Orogeny had somewhat matching effects and was largely responsible for the nature and distribution of different facies of marine and continental rocks in the Appalachian area.

sands and calcarenites; carbonate, evaporite and organic sedimentation was widespread and persistent.

Carbonate and evaporite deposition had begun in Silurian time. Devonian facies were much the same, with locally impressive Middle and Upper Devonian algal, stromatoporoid and coral reef developments as the Kaskaskia Sequence progressed (see figure 7.10). During Middle Devonian time most of the Canadian Shield was submerged. Small ecologic reefs occur east of the Transcontinental Arch but in the western Canadian (Willison and other) Basins ecologic reef growth was luxuriant (see p. 150). Late Devonian black shale covers much of the carbonates east of the Transcontinental Arch. Little detrital influx occurred and these conditions persisted with the Lower Carboniferous.

West of the Transcontinental Arch also the Lower Devonian series, where present, is thin, while the Middle and Upper Devonian carbonates and evaporites constitute the major part of the transgressive sequence. As far south as Nevada the earliest Devonian is a regressive succession and in the vicinity of the Rocky Mountains a low land mass existed south of the 49th parallel. Much detailed work has taken place here since the Calgary Symposium (Oswald, 1968; see Mallory *et al.*, 1972, also Murphy *et al.*, 1977). Near the western edge of the craton in the U.S.A. the late Frasnian Antler orogeny is indicated by the Roberts Mountain Thrust and other structures that face generally eastwards.

Detailed biostratigraphic successions in shallow-water carbonates have been worked out in several areas of the U.S.A. Facies changes are significant in the region and a Lower Devonian near-shore facies in several states has yielded a rich vertebrate fauna. Hercynian stage names have been used in correlations that were made on the basis of brachiopods and conodonts. Late Devonian vertebrates are also known from near-shore carbonate units with marine invertebrates. As in Germany, eastern North America and elsewhere, contemporaneous brachiopod communities have been distinguished in parts of this shelf succession. They seem to depend upon the usual factors of depth of water, substrate and agitation of water.

Murphy *et al.* (1977) and others have by careful biostratigraphy been able to show the minor regressions and onlaps that occurred during Late Devonian times in the west, and have noted the widespread extent of Middle Famennian emergence of much of the craton there.

Figure 7.11 shows some stages of these in the evolution of the North American continental interior.

Epicontinental Successions outside North America

North Africa

Devonian rocks are widespread across the great North African craton, from southern Morocco in the west to Egypt in the east (see figure 7.11). As in North America, flexures and fractures of the basement have influenced sediment accumulation and the siting of reefs and biostromes. Terrigenous sediments are generally fine to medium grade, disposed in thin units over wide areas. The carbonates include a wide variety of types that range from *Orthoceras*-rich limestones to true reefal accumulations of algal, stromatoporoid and coral limestones. Cyclic or repetitive sedimentation of thin clastics and carbonates over wide areas occurred in Early Devonian time (Hollard, 1968). The seas seem to have achieved their most widespread extent in the Givetian, the phase also perhaps of the most extensive reef growth. The Moroccan formations in particular appear to be highly fossiliferous with pelagic faunas, especially ammonoids, well represented in many formations. In the Hamar Laghdad area Early Devonian tuffs provided a firm substrate upon which extensive reefs developed in Emsian time. East of Morocco (Algeria, Tunisia) the Devonian outcrops are very limited but the system is encountered in many boreholes (see figure 7.12). As in the west, most parts of the system are represented by highly fossiliferous strata. The lithological succession is well known, but again as in Morocco, precise chronostratigraphic detail is locally uncertain. Carbonates are almost entirely restricted to the Middle Devonian and at the top of the system a regressive sandstone phase is present. Many formations may be diachronous to some degree. The origin of the clastic sediment is largely to be found in the cratonic area to the south-east (Vos, 1981). In the Koufra basin of southern Libya the entire

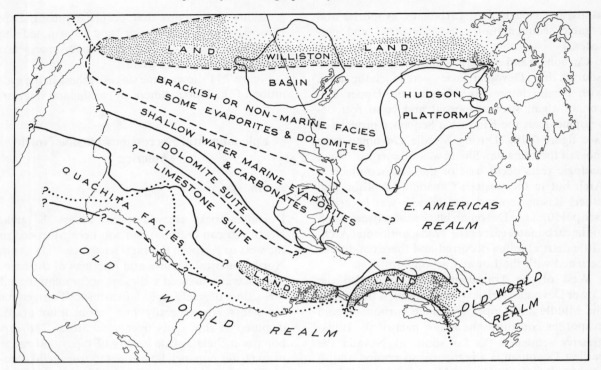

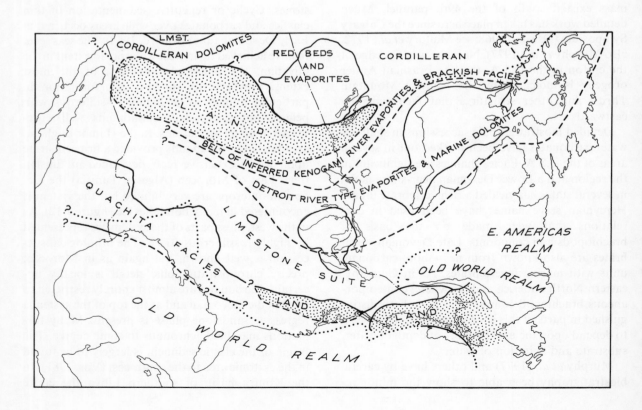

Devonian succession is clastic, being composed of fine sandstone and siltstones and attaining a thickness of about 250 m.

Shallow-water Devonian strata extend even farther eastward on to the Arabian Shield where they occur as the poorly known Tabuk Formation in the north-central area. Devonian formation of Old Red Sandstone or shallow-water marine facies extend in a latitudinal belt across Iran, northern Afghanistan and Pakistan. Everywhere here they appear to be transgressive upon much older units. Thus the later stages of the Devonian are more widespread than those of the Lower Devonian. Middle Devonian reefal carbonates occur in central Afghanistan and south of this entire region geosynclinal facies are developed. This state of affairs persists eastward

into Pakistan and India with carbonates and quartzites being deposited on a very shallow marine platform.

Russian Platform

The Russian (see figure 7.13) and Siberian Platforms constitute two huge areas of widespread Devonian deposits that include a very wide range of relatively shallow-water lithologies and faunas. In addition to these regions, many of the marginal or shelf zones that border the Ural, Caucasus and other basins reveal well-preserved Devonian successions. Marine and continental facies are contemporaneous or alternate in many areas and

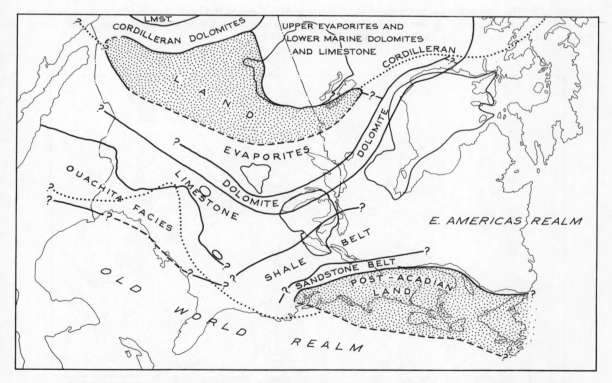

Figure 7.11. Three stages in the evolution of the North American continental interior (Upper left, Gedinnian; lower left, Eifelian; lower right, Givetian) show facies belts and land masses largely orientated latitudinally (after Boucot, 1975). Here the Eastern American Realm was confined on the south by land masses and the shale–novaculite Ouachita facies and on the north by the Transcontinental Arch and hypersaline environments extending from California to northern Canada. The faunas of the Cordilleran Region (lower left and lower right) belong to the Old World Realm (faunal realm boundaries are shown as dotted lines) and it is interesting to note there is apparently greater faunal affinity between that subprovince of the Old World Realm and that of eastern Asia than with the perhaps contiguous Eastern American Realm so effective were the Williston land and hypersaline environments as barriers to migration.

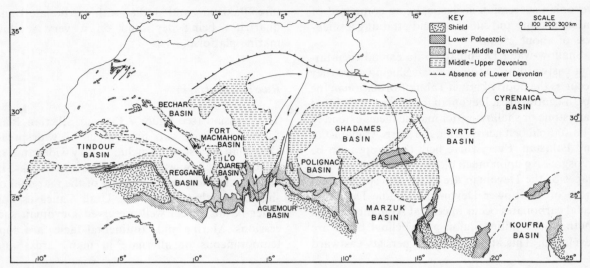

Figure 7.12. The Devonian epicontinental seas in North Africa deposited both clastics and carbonates in a general facies distribution that was influenced by the development of broad shallow basins and intervening narrow uplifts (after Hollard, 1968; Legrande, 1968; and others).

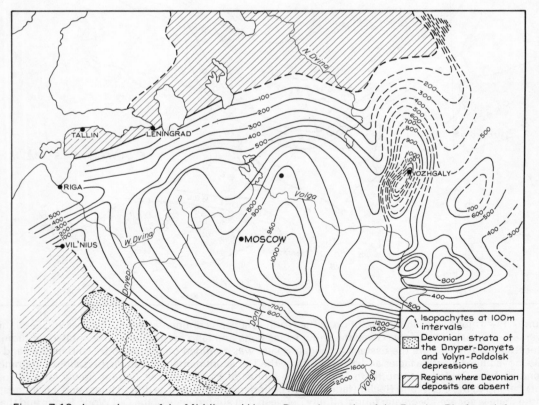

Figure 7.13. Isopach map of the Middle and Upper Devonian rocks of the Russian Platform (after Nalivkin, 1973a, 1973b). To the north, north-west and south-west are basement areas from which much clastic sediment was derived; to the east the platform gave way to the Urals 'geosyncline'.

clastic, carbonate and volcanic rocks may all be present. The column in the Russian Platform reaches a maximum thickness of about 2000 m depending upon the local conditions (Sokolov *et al.*, 1981). In this platform, as elsewhere, separate basins and arches, fault-controlled or not, are a conspicuous feature of the geology. An enormous wealth of data has accumulated from the wide subsurface exploration of this region in recent decades, especially from the search for oil and gas (Lobanowski and Przybylowicz, 1979).

All parts of the system are represented in the Russian Platform with facies ranging from deltaic and alluvial, through lagoonal and evaporite to bituminous limestones (Domanik facies, see page 167, and deeper-water, outer-shelf types of shale. Source areas for the sediment lay in the Baltic Shield and in arches and domes within the limits of the platform itself.

Transgression seems primarily to have been north-westward and westward towards the Baltic land mass: during regressive phases the sea withdrew towards the Ural Ocean.

Early Devonian deposits are thin, terrigenous and discontinuous; much of the area was above base level. Middle Devonian sediments are thin but widespread, resulting from a wide transgression which gave rise first to lagoonal and shallow nearshore shelf conditions where salinity was locally higher than normal, and then to normal salinity and open shelf conditions. Argillaceous and carbonate rocks are widespread. A phase of unrest in Givetian time is indicated by local uplift and a heavy influx of terrigenous material. Transgressive and regressive successions alternate over much of the platform and volcanic activity in the north and in the south may indicate crustal fracturing. Late Devonian formations are of widespread red terrigenous clastics in the west and lagoonal to open-water marine shales and carbonates (and Domanik facies) in the east. Frasnian strata may reach a thickness of up to 1000 m but Famennian rocks are generally less than 250 m thick, evaporitic and clastic, and indicate a gradual rise of the platform.

Asia

The Siberian Platform is the geological heartland of north-eastern Eurasia, stretching from the Arctic Ocean to Lake Baikal and from the Yenisei River to the Lena and Aldan. Much of the platform was the site of non-marine deposition in Early Devonian time but in the Middle and Late parts of the period shallow-water lagoonal clastics, evaporites and carbonates were widespread along the margins of the platform. Devonian volcanic rocks occur at many places (see figure 7.14). The marine faunas include abundant brachiopods and molluscs; coelenterates are more particularly common in the higher units. At the end of Ludlow time the platform was above sea level and the sea only began to transgress at the end of Early Devonian time. Thereupon it spread rapidly from the east into most areas, except the southern part of the platform where the Sayan Mountains were rising. The Middle Devonian saw an early regression and erosion, followed eventually by a Late Givetian extensive transgression. In the Frasnian a renewed regression exposed all but the northern parts of the platform surface but it was overcome by the most widespread flooding. Even under these conditions higher than normal salinities appear to have prevailed over much of this huge area. Towards the end of the Famennian a regressive tendency was once again established but volcanic activity was intense and prolonged in the south-central and eastern-most parts of the platform. Soviet workers distinguish a number of structural facies regions within the platform which reflect the tectonic behaviour of the platform during this period (see figure 4.8).

China

Continental facies give way laterally to littoral and 'platform' facies in several parts of central and southern China (Yang *et al.*, 1981). Chinese stratigraphers distinguish two types of magnafacies in these regions. The Xiangxhou type is made up of bioclastic limestone, dolomites and shales that are interbedded with terrigenous clastics and with an abundant fauna, while the Nandan type also includes platform sandstones and black mudstones. This latter facies is of a more offshore and deeper-water origin, being situated on the outer parts of the broad shelf. Most of the South China region was covered by a shallow sea throughout the Devonian

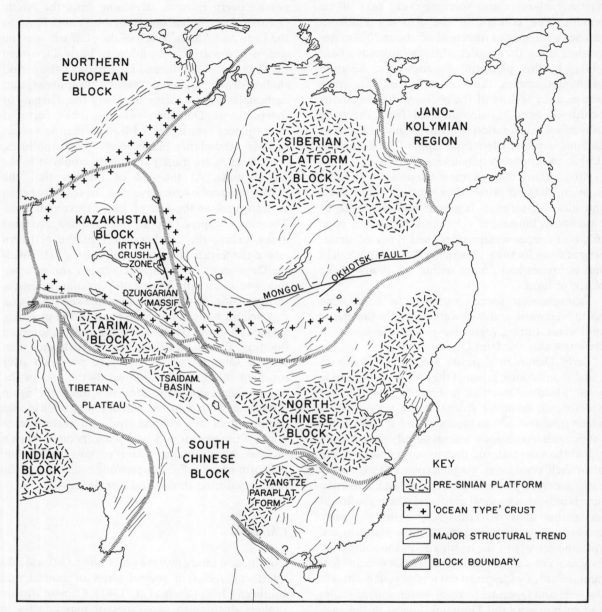

Figure 7.14. A tectonic map of part of east Asia (Burrett, 1974) shows cratonic blocks over which epeiric sediments were spread in Middle and Late Pelaeozoic times, and 'mobile belts' between in which geosynclinal and post-orogenic sedimentation occurred. Caledonian, Hercynian and Alpine orogenies and ocean closures have brought about this unification, especially conspicuous being those of the Uralian, Angaran and central Asian geosynclinal regions.

and all parts of the system are represented there, although the total thickness is commonly less than 2200 m. Terrestrial intercalations are most frequent in the Lower Devonian; the Middle Devonian possesses few such beds. Middle Devonian carbonates are highly fossiliferous with large brachiopod and coral assemblages. The Upper Devonian, especially in Kwangsi Province, contains highly fossiliferous limestones and mudstones that are indicative of widespread and relatively uniform open shelf environments. The relatively large palaeogeographical changes that occurred in central and southern China throughout the Devonian period seem to indicate vertical crustal movements that are comparable to those in the other great cratonic areas described above. Geosynclinal seas virtually surrounded the South China cratonic area and some were sites of much contemporaneous volcanic and tectonic activity. Research into the geological evolution of this area by Chinese geologists has been intensified in recent years and the results are increasingly being communicated to the West.

Australia

In central and Western Australia the wide distribution of Devonian rocks suggests that much of the craton, bordered in the east by the Tasman Geosyncline, was eventually covered by this system (see figure 7.15). The major existing outcrops occur in local cratonic basins, the Carnarvon, Canning, Bonaparte Gulf and the Amadeus and Ngalia Basins, in all an area of more than four million square kilometres. Almost all rest unconformably upon Lower Palaeozoic or Precambrian rocks. Early Devonian strata are primarily terrigeneous, shelf sandstones with abundant sedimentary and biogenic structures. They may reach a maximum thickness of 1000 m in the Canning Basin. The Middle Devonian is primarily of carbonates with biostromal units in the Carnarvon Basin and reefal limestones from the Middle to Upper Devonian of the Canning and Carnarvon Basins are justly famous (see chapter 8). Over some 280 km along the north-eastern margin of the Canning Basin the facies, structure and evolution of a series of fringing reefs and 'atolls' has been studied (Playford, 1980).

A maximum thickness of over 1000 m of basin deposits occurs within the Canning Basin, accumulated in a fault-controlled area and derived from the east and south.

The Bonaparte Gulf Basin contains Upper and Middle Devonian dominantly terrigenous clastic and fossiliferous beds, in all 3000 m thick. The Amadeus Basin in central Australia is also filled by about 5000 m of siltstone and coarser clastics. Some authorities have regarded the flooding of the Australian craton as being confined to relatively narrow channels and gulfs (Johnstone *et al.*, 1968); others have suggested a wider spread to the epicontinental seas (see figures 6.15 and 7.15).

South America

Very large shallow sedimentary basins covered much of the craton of South America during the Devonian. They appear to have been large embayments opening from the ocean to the west and to have been marked by very migratory shorelines. On occasion they extended beyond South America on to the adjacent surface of the African continental plate, as in Ghana. Today open shelf sequences are known in most parts of the Andean region and the basins of the Brazil–Paraguay–Uruguay region are known as the Amazon, Parnaiba, Tucano, Jatoba and Parana Basins (Copper, 1977; see figure 7.16). They belong to the Gama cratonic sequence Lower Cambrian–Silurian) (Mabesoone *et al.*, 1981).

The sediments deposited in these basins were entirely clastic, and detrital, ranging from black muds to coarse quartz sands: no carbonates are known. Volcanic rocks are also entirely absent. The thickest accumulations are around 600 m in the Parana Basin (see figure 7.17) and 1400 m in the Amazon Basin. Most, though not all, of the sediments were derived from the north and east. There has been very little tectonic disturbance and virtually all the formations are flat-lying and undeformed. The absence of carbonate, the rather restricted kinds of fauna present and palaeomagnetic data suggest high southern latitudes and a cool-water to cold-water regime. Sessile filter-feeding coralline animals were extremely rare or absent and the numbers of other taxa such as

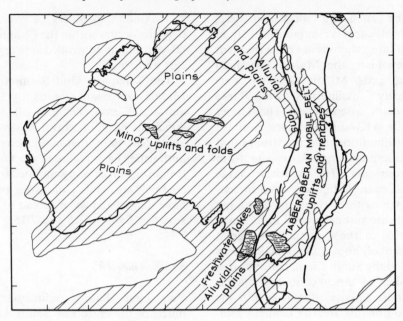

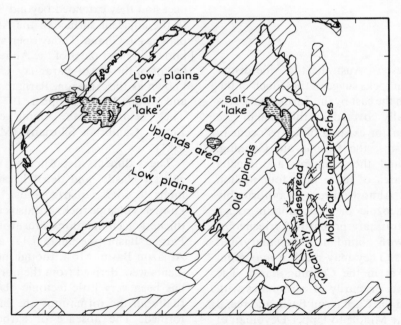

Figure 7.15. Australian palaeogeography. While the greater part of the continent was of low relief during Devonian time a central upland belt shed sediment to the north and south and the active mobile zone of the present east coast was the scene of great volcanicity, linear uplifts and subsidence on the grand scale (after Laseron, 1969; and others). Top—Early Devonian; Bottom—Late Devonian.

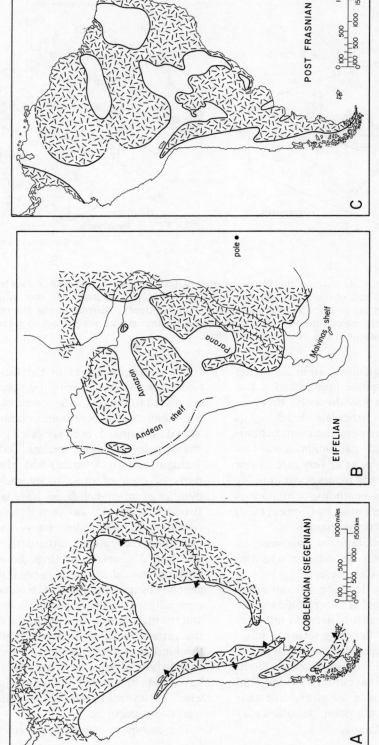

Figure 7.16. The changing Devonian geography of South America (largely after Copper, 1977). Land areas jackstrawed, arrows show provenance of sediment. To the 'west' of the Andean and Malvinas shelves was an active continental margin, part of the Samfrau mobile belt (geosyncline). The large and shallow continental–epeiric basins were sites of dominantly clastic shallow deposition.

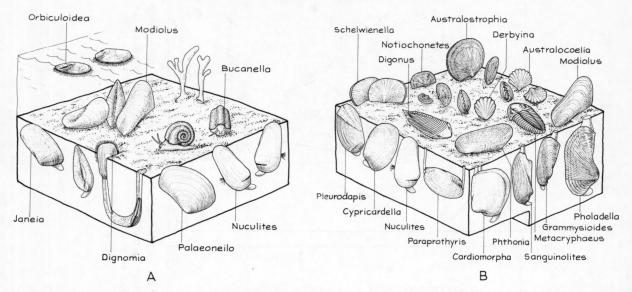

Figure 7.17. Reconstruction of cold-climate (Malvinokaffric realm) shallow-water communities from the Emsian of the Parana Basin of Brazil (largely after Copper, 1977). (A) Marine to intertidal: the lingulid *Dignomia* community of the Jaguariaiva Shale. (B) Shallow marine (to 50 m depth) *Australocoelia* community from the Jaguariaiva Shale, a very diverse community that contained many infaunal bivalves but lacked the major groups of brachiopods that were present in the other marine realms.

brachiopods were much smaller than in the higher latitudes. In short, the sparse faunas had a low population density and a low diversity. Bivalves, inarticulate and certain articulate brachiopods dominated, while trilobites were common. Echinodermata, gastropods and nautiloids were rare and coral-like forms are absent or very rare. A few shark remains are known and plants occur at many levels. Studies of the local benthonic communities are at an early stage. Pioneer work by Copper (1977) has demonstrated their general characters.

Out on the Andean Shelf, bryozoa, rugose corals and crinoids were locally in abundance, and ammonoids and conodonts were present.

The highly changeable location of shorelines seems to reflect the low relief of the cratonic surface and the gentle warping that it underwent throughout the period. This mild flexuring seems to have been rather more conspicuous to the south where the Pampean Fold Belt was to evolve. Slight changes of sea level would have effected widespread transgression or regression of the strand-line as is indicated by the maps. South polar glaciation may have been an additional factor.

The enormous extent of the Devonian epicontinental seas together with the open waters of the oceans led to the very wide spread of marine animal communities. The benthonic communities have, as stated in chapter 5, been allocated generally to three major palaeobiological realms, Old World, Appalachian (eastern America) and Malvinokaffric. In parts of each of these outer shelf facies and a number of inner shelf facies have been recognised. Broad relationships between different brachiopod communities and sedimentary environments are becoming recognised, particularly in the Middle and Late Devonian. By then, however, the three biogeographical realms were breaking down into a general cosmopolitan state. Carbonate shelves were especially widely developed in the tropical latitudes and the high species diversity of the time may reflect the expansion of shallow-water environments of this kind during later Devonian times. The extreme shallowness of some very large areas of water may have made many marine faunas highly vulnerable when, as towards the end of Devonian time, sudden regressions occurred.

The causes of sea-level movement remain uncertain but the major control must have been orogenic—a response to the Caledonian, Acadian, Antler and other orogenies. Oceanic ridge growth and decline would also have operated an eustatic control. In addition, climatic oscillation, land-based erosional cycles and local isostatic activity would perhaps have played a part. As biostratigraphic correlation improves, the precise timing of these movements of sea level and their accompanying facies belts will become apparent. Rates of movement may be better calculated and the underlying mechanism may be indicated more clearly.

8

Warm Waters, Carbonates and Evaporites

Amongst the major lithologies present within the Devonian assemblage (figure 2.1, page 21) there is a remarkable swing in the proportion of the carbonates present. An all-time low in the Lower Devonian is followed by an increase to about 10 per cent. Mixed carbonate and clastic deposition, however, rose from about 20 per cent to over 30 per cent during the same time. The increase appears to be largely at the expense of marine clastics. The change, accompanied by diversification of the

marine faunas, was in keeping with world geographical changes. In this chapter we briefly survey some aspects of carbonate and evaporite deposition and their distribution within, and their importance for, the Devonian world (see figure 8.7).

While we may regard the clastic rocks as the product of physical processes of sedimentation, the carbonates and evaporites (the bulk of the remaining sedimentary array) are largely the product of chemical processes, or perhaps more immediately in

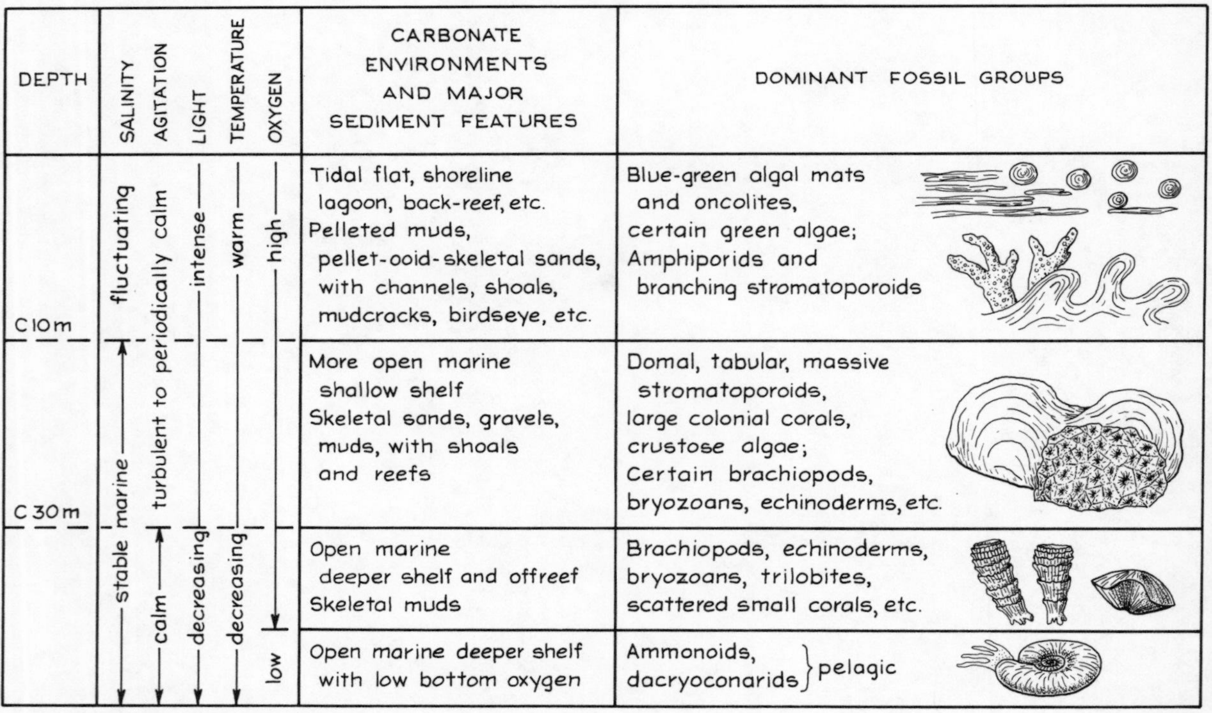

DEPTH	SALINITY	AGITATION	LIGHT	TEMPERATURE	OXYGEN	CARBONATE ENVIRONMENTS AND MAJOR SEDIMENT FEATURES	DOMINANT FOSSIL GROUPS
C 10 m	fluctuating / stable marine	turbulent to periodically calm / calm	intense / decreasing	warm / decreasing	high / low	Tidal flat, shoreline lagoon, back-reef, etc. Pelleted muds, pellet-ooid-skeletal sands, with channels, shoals, mudcracks, birdseye, etc.	Blue-green algal mats and oncolites, certain green algae; Amphiporids and branching stromatoporoids
C 30 m						More open marine shallow shelf Skeletal sands, gravels, muds, with shoals and reefs	Domal, tabular, massive stromatoporoids, large colonial corals, crustose algae; Certain brachiopods, bryozoans, echinoderms, etc.
						Open marine deeper shelf and offreef Skeletal muds	Brachiopods, echinoderms, bryozoans, trilobites, scattered small corals, etc.
						Open marine deeper shelf with low bottom oxygen	Ammonoids, dacryoconarids } pelagic

Figure 8.1. The vertical distribution of carbonate environments and the major fossil groups that contribute to sedimentation reflect response to the factors shown on the left. These factors tend to vary with water depth in warm seas (after Heckel and Witzke, 1979).

the case of carbonates, biological processes. In practice it is often difficult to separate carbonate stratigraphy from studies of the accompanying evaporites. Here it is convenient to treat carbonates first and follow this with a consideration of evaporites. Carbonate formations extend far back into the Proterozoic but became conspicuous to an unprecedented degree early in the Phanerozoic. Carbonate production is basically an organic activity and the more important lime-secreting organisms are very sensitive to environmental conditions. Their presence in the stratigraphic column is taken to imply similar conditions in the past and thick carbonates may be regarded as indications of rather long-enduring suitable habitats for these organisms. Given an understanding of the ecology of the organisms in question, much palaeogeographical information (climatic, hydrographic, bathymetric and biological) can be retrieved. Conditions may be inferred with a fair degree of confidence, as may also sedimentational (figure 8.1) and palaeotectonic factors that influence depositional changes in sea level on a local or world-wide scale. Johannes Walther, known for his 'Law of facies' was, incidentally, amongst the first to advocate an actualistic approach to ancient reef and carbonate formations (see Middleton, V. W., 1973).

Carbonate-forming and lime-producing organisms include coelenterates, sponges, coralline red algae, blue-green algae, bryozoans and molluscs. Frameworks (skeletons), encrustations and an abundance of fragmental particles, faecal pellets and other calcareous matter all contribute to carbonate sediments past and present. Different organisms have filled these roles throughout geological time but their overall effect has varied remarkably little over the last 400 million years. Algae are regarded as paramount in the role of binding other organic materials in the carbonate, and especially the reef, environments. Several types of calcareous algae have this function in the Devonian (see Riding, 1979)—figure 8.2. They are also very important in the restricted shallow-water marine environments where they occur as skeletal or non-skeletal stromatolites, laminites, oncolites, or as loose fragments (Tsien, 1979). Identification is difficult and the products of algal growth and their subsequent decay are not always readily determinable. The reef-building or binding activity of the supposed blue-green alga *Renalcis* has been highlighted in Western Australia (Playford, 1968), Canada (Mountjoy and Riding, 1981) and Belgium (Tsien, 1979). Other important Devonian genera are *Epiphyton, Girvanella, Solenopora* and *Rothpetzella*.

It is important to recognise that marine carbonate sediments today accumulate in two distinct

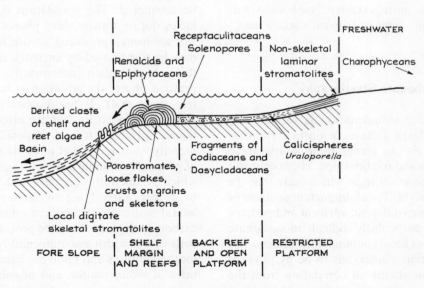

Figure 8.2. Common associations of algae in shallow-water marine Devonian environments (after Riding, 1979).

kinds of environments. Shallow-water carbonate environments are essentially tropical, being confined to continental shelves, arcs and submerged platforms and volcanic horsts and stumps. True oceanic carbonates are deep-water sediments, which accumulate under quite different conditions. Oceanic sediments occur in rather minor proportions in the stratigraphic column and are not especially noted for their carbonate content.

Local conditions for the active production of shallow-water carbonate sediments today include normal pH value and salinity, temperatures of around 21°C, depth of generally less than 100 m, little turbidity, and constant agitation, although not great disturbance, of the water. These conditions are most abundantly developed where coral reefs and carbonate shelves occur in the tropical latitudes in the Indo-pacific realm, the Antillean realm and on the west coast of Mexico. They are easily accessible for study and provide models of several kinds for comparison with ancient carbonates. In a general account of carbonate facies throughout geological history, Wilson (1975) points out that carbonate deposition is rapid but easily inhibited, and therefore it is sporadic during geological time. It is also one of the peculiarities of carbonate rocks that they are subject to many stages of diagenesis which may hinder our interpretations of their origins. Despite these difficulties it seems possible to regard virtually most extensive carbonate formations as the products of tropical shallow seas.

Carbonate Stratigraphy

Because carbonate sedimentation is so distinctly different from clastic sedimentation in certain respects, it gives rise to recognisable stratigraphic patterns of facies and relationships. They occur in a variety of tectonic settings which may not be apparent at first sight. The all-important problem of correlation between different, adjacent and perhaps coeval facies is particularly difficult in carbonate suites whose facies fossils dominate the faunas. The recognition of 'time planes' has to be by physical criteria, or by an attempted correlation from the

ecological relationships of the facies fossils, or by a reconstruction of their ecosystem.

The shapes of carbonate bodies may be highly distinctive and recognisable immediately from subsurface geophysical data. As in other areas of sedimentology the terminology for these lithosomes has been extended and it is important to recall that classifications and definitions may refer to shapes, internal compositions, sizes and, perhaps prematurely, genesis. In Wilson's (1975) useful account of carbonate-formation terminology the term '*carbonate buildup*' denotes a body of locally formed (laterally restricted) carbonate sediment that possesses topographic relief (figure 8.31). '*Reef*' is a term that is likely to give trouble as it has been used in so many different ways (Heckel, 1974), so it is best used with a modifying word to indicate its meaning. Invariably it refers to a local mound-like body that is distinct from adjacent facies. As James (1977) remarks, 'The way to understanding reef facies is unravelling the complex series of lithologies that comprise the reef core.'

The basic facies pattern for carbonates commonly reflects environments that are situated well away from terrigenous contamination, in the shallow (warm) waters of shelf seas and tectonically stable areas. The correlation between different facies during tectonically stable phases is particularly difficult. An ecostratigraphic approach must be attempted. The correlation between different facies during transgressive phases is much easier. The sediment is produced 'on site' by organisms but may be transported by currents, tides and gravity flows, landwards or oceanwards. During a transgressive phase, the situation is however different (see figure 8.4). It has become the practice to recognise three belts that are associated with carbonate production and deposition. These are broadly 'back reef' (that is, to landward of the site of greatest carbonate production), 'shelf margin' at which production is greatest, and 'basin' (the area of deeper water lying to seaward). A standard lateral sequence or pattern of facies belts for carbonate deposition can be postulated. The facies belts vary in width and uniformity, depending upon local conditions and upon the balance between the rates of sedimentation and of subsidence. Thick-

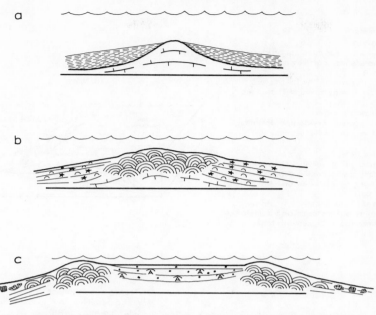

Figure 8.3. Middle Palaeozoic buildups were commonly produced by vigorously growing organisms that were able to keep pace with subsiding substrates. (a) A micrite mound that contains bryozoa with algae and/or sponges; generally not very large but symmetrical and surrounded by shale. (b) A patch-reef with central stromatoporoid zone, flanked by crinoidal debris. Shapes may be round or elongate, trending parallel to contours of substrate. (c) Banks of low relief with stromato-poroid rims and amphipora (A) spreads in the interior and tabulate and colonial corals on the outer flanks. These range from giant banks, hundreds of kilometres long, to circular atolls, less than 1 km wide (after Wilson, 1975).

nesses also depend upon these factors; prograding and regressive patterns of carbonate generation are well known (figure 8.4), and repetitive or cyclic sedimentation fills basins as a result of alternating sea levels or tectonic activity. The study of Holocene carbonate environments has helped in the development of a single simple (ideal) model for carbonate facies (see figure 8.5).

Topographic, tectonic control of depositional sites has frequently resulted in linear trends of carbonate buildups on shelf edges and within basins, and tectonic activity leading to erosion and/or regeneration of buildups may afford criteria for interfacies correlation. Eustatic changes of sea level greatly influence carbonate growth and may result from world-wide glaciation or changes in ocean ridge and trench volumes. Temporal emergence of carbonates can be used in correlation.

Diagenetic changes almost invariably result from sea-level fluctuations, and while they may be of some limited use in correlation they also tend to destroy stratigraphically important fossils as well as other organic structures.

The tectonic settings most conducive to carbonate accumulation on the grand scale are those of stability or slow changes of level. Areas of orogenic uplift tend to shed terrigenous sediment into surrounding seas and thus to inhibit carbonate organisms. The following general outline of relationships, tectonic settings and major categories of carbonate suites is basically that which Wilson (1975) adapted from previous authors.

A. Relatively stable phase
  I. On the stable platform
    1. Biostrome (R4): widespread tabulate or layered reef, mainly coverstones, bafflestones, bindstones and framestones.
    2. Patch reef (R2): isolated in the restricted carbonate, mainly framestones, bafflestones and bindstones.
  II At a shelf or platform margin
    3. Barrier reef (R1): coarsely crystalline dolomite, mainly framestones and coverstones.
  III Within a basin
    4. Bioherm complex (R3)
      a. Type Arche reef (R3a): vertical then horizontal variation: mainly coverstones and biocementstones in the lower part, coverstones and framestones in the upper part.
      b. Type Lion reef (R3b): mainly horizontal variation: mainly bafflestones and bindstones in the peripheral part, biocementstones in the central part.

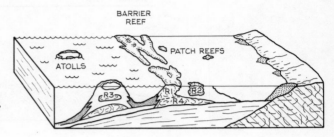

B. Transgressive phase
  IV. In a basin
    5. Mud mound (R5)
      a. Type Neuville reef (R5a): mainly lamellar corals and algae (bindstones and biocementstones).
      b. Type Beauchateau reef (R5b): mainly branching corals and algae (bafflestones and biocementstones).
      c. Type Fort Condé mound (R5c): mainly biocementstones with *Stromatactis* and sponges.
  V. Fringing a landmass
    6. Fringing reef (R6): mainly coverstones and framestones

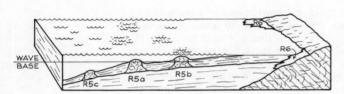

Figure 8.4. A synopsis of the classic Belgian reefs and bioherms by Tsien (1979) relates reef form and carbonate components to position during stable or transgressive phases of sea level. Variations in sea-level behaviour, together with short phases of marine regression, are thus held to be prime factors in the pattern of reef growth here, as elsewhere in the world and throughout Phanerozoic time.

1. Basinal buildups in areas of major subsidence.
   a. In marginal cratonic basins or miogeosynclines; for example, Leduc and associated Upper Devonian buildups, central Alberta Basin, Williston Basin, western Canada.
   b. Within geosynclinal troughs or on or around volcanic uplifts; for example, Rhenish Schiefergebirge Middle Devonian.
   c. Major offshore banks under oceanic influence; for example, Lion reef, Dinant Basin, Belgium.

2. Buildups at edges of major platforms and ramps that developed off cratonic blocks: areas of major subsidence.
   a. Linear buildups along margins of these platforms; for example, Presqu'ile and around Peace River high, northern Alberta, Canada, and the Lennard Shelf of north-west Australia.
   b. Complexes of individual banks and reefs that are widely spread along and upon such shelf margins; for example, the Eifel area of West Germany, and in south-west Morocco.

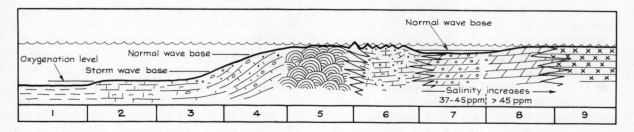

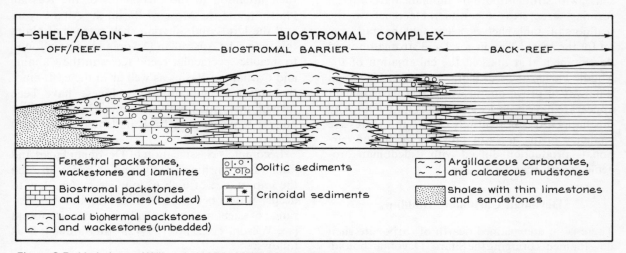

Figure 8.5. Variations of Wilson's (1970–1975) idealised sequence of facies belts (above) (see Wilson, 1975) is to be seen in many Devonian reefal carbonate areas such as that of the Portilla Formation of the Cantabrian Mountains in northern Spain (below). 1. *Basin Facies*—largely dark argillaceous and siliceous sediments. 2, *Open Shelf facies*—well below wave base carbonates and argillites, very fossiliferous. 3, *Toe of slope carbonates*—fine-grained, with bioclastic detritus. 4, *Foreslope*—bedded and slumped carbonates and shales, very fossiliferous. 5, *Organic (ecologic) reef*—marine organic structure, rich in carbonates. 6, *Back reef face*—lime sands and oolites. 7, *Open platform*—variable clastics and carbonates, restricted marine fauna. 8, *Restricted platform*—lagoonal muds etc., lagoonal faunas. 9, *Platform flats or sabhka*—evaporites, caliches, red beds, generally barren. The Portilla (Middle Devonian) section (after Burchette, 1981) is about 2 km long, and the biohermal masses are about 10 m thick.

c. Major but narrow fringes off faulted edges of cratonic blocks or orogenic ridges.

3. Buildups on ramps or platforms in areas of moderate subsidence; buildout into shallow intracratonic basins; on shelf areas with scattered mounds and patch reefs.

    a. Micritic mounds and masses with very slight relief in shallow water, ordinarily with only sponges or bryozoa and an algal/stromatolite cap.

    b. Simple mud-mound accumulations with calcarenite flank beds.

c. Roundstone reefs atop banks or mud mounds; for example, the Devonian patch reefs of Spanish Sahara (south Morocco).

d. Faros or patch reefs of pure boundstone; for example, the Devonian over wide area of North America, and the Ardenne–Eifel area of Europe.

e. Upward shoaling cycles across ramps from shelf to basin.

4 and 5. Low fringing buildups or haloes of cyclic character around local positive areas on shelves.

In the Early and Middle Cambrian the mound-building activities of the calcareous algae were accompanied by the development of archaeocyathid reef structures and by mid-Ordovician times the lime-precipitating coelenterates were active. The Ordovician Chazy carbonate mounds reveal algae, bryozoa, stromatoporoids and pelmatozoa, which jointly contributed carbonate materials. Silurian time saw the advent of reefs in a real sense, with stromatoporoids, tabulata and corals all forming structures of considerable relief on the seafloors of stable shelves. Seemingly the stage was set for the great boom in coral and stromatoporoid development that marked the culmination of the Middle Palaeozoic reef biota. Yet Early Devonian time was to prove one in which reef development appears to have been geographically inhibited. Why this happened is not entirely clear but a regressive tendency in the areas of previous maximum carbonate production, linked with Caledonian orogenic events, may be suspected.

Devonian Carbonate Buildups

There is an unexplained dearth of carbonate shelf developments bridging the Siluro–Devonian boundary. Where carbonate facies do occur they consist of those associated with coral and stromatoporoid patch reefs, like, in some aspects, those known from the American Midwest and the Baltic areas.

In Middle and Late Devonian times, however, a world-wide burst of evolution overtook the corals and stromatoporoids. It was a highly significant process. With the rapid spread and growth of the corals and stromatoporoids over enormous areas of epicontinental seafloors and continental margins, carbonate buildups soon took on many shapes and occurred under an increasing range of environmental circumstances. Not only were isolated mounds and patches abundant but in several areas large linear barrier reefs grew up.

One of the contributing organisms to influence the establishment of new reefal masses was *Thamnopora*. It is widespread throughout the Devonian but in the remarkable Siluro–Devonian reef complex in Pakistan described by Stauffer (1968) its role was vital. From a base of poorly fossiliferous micritic limestone there developed a layer of *Thamnopora* colonies and other framework builders.

Undoubtedly the most influential palaeoecological studies of the carbonate and reef facies of the Middle and Later Devonian have been those of M. Lecompte and H. H. Tsien in the Dinant area of Belgium. Work on western Canadian carbonates began in earnest in the early-1950s with the discovery of petroleum in reef limestones in the subsurface there: it subsequently extended to the Canadian Arctic Islands. Soviet geologists turned their attention to the carbonates of the Russian Platform and the Moscow Basin, the Altai-Sayan, and the Urals and Siberian shelf regions at about this time. The Australian Devonian has been found to include spectacular reefal facies in the Canning and Carnarvon Basins as well as in the east. Since the cultural revolution the Chinese have been turning their attention to the great Devonian carbonate sequences in the Karstlands of southern China.

Lecompte's classification of Devonian facies and faunal relationships (see figure 8.6), which stressed the importance of high to low energy (shallow to deeper water) beds, seems to encompass the great range of carbonate rocks wherever they are found (see Wilson, 1975, pp. 120–125). In outline it is as follows:

Basinal facies

1. ('Deep zone'). Brown-black, organic-rich shale with dacryconarids, *Buchiola*, conodonts and goniatites. Deposits around buildups in deep, poorly aerated waters; Duverney of Alberta and inter-reef deposits of Franco–Belgian area, western Australia etc. Water as much as several hundred metres deep.

2. Grey-green shale with ostracods, dacryconarids, *Buchiola* and some foraminifera. Ireton shales of Alberta. Water perhaps more than 100 m deep.

3. ('Quiescent zone'). Normal marine lime wackestone–packstone, may be brown or cream colour. Varied form dominated by brachiopods (excluding rhynchonellids), scattered rugosa and bryozoa, abundant bioclastic debris. Water less than 100 m.
   Facies 1 to 3 listed here are absent from the near-shore and back reef environments where lagoonal and sublagoonal conditions prevail.

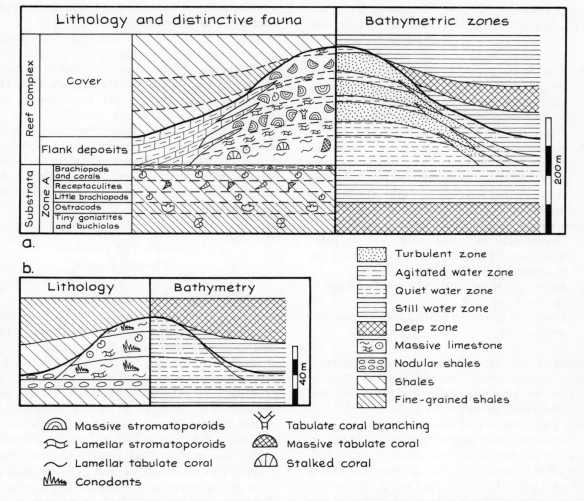

Figure 8.6. The ecological associations and bathymetric zones of Devonian bioherms in the Franco-Belgian Devonian, as suggested by Lecompte (1970) in a series of remarkable investigations.

Shelf or platforms with open circulation

4. Small, globular stromatoporoid-bearing shelf wackestone–packstone in biostromal layers with a few corals and red algae. Normal marine limestone fauna. Formed just below wave base in very shallow water.

Gentle slopes on banks

5. ('Below turbulence zone'). Coral beds. Large heads of dendroid and fasciculate rugose corals (*Disphyllum*) and massive forms (*Phillipsastraea*), dendroid stromatoporoids (*Stachyoides*). Bioclastic wackestone. Low on slope (=coral bafflestone of Embry and Klovan, 1971). Water depth perhaps 20–30 m.

6. ('Subturbulent zone'). Tabular stromatoporoids (25–50 m) with much *Stachyoids* (dendroid stromatoporoids) and angular bioclasts. Biota normal marine, varied with brachiopods and crinoids. Mid-slope limestone facies, lacking clay mud. May have coral beds. Depth perhaps 10–20 m.

7. *Stachyoides* thickets. Boundstone with many irregular dendroid forms and some tabular stromatoporoids, brachiopods, crinoids and red algae.

Bank or shelf margin facies

8. ('Zone of turbulence'). Marine irregular stromatoporoids. Middle of margin slope with

grainstone–packstone matrix; much *Stachyoides* red algae and normal marine bioclasts. Fairly agitated water. Water only a few metres deep.

9. Massive-irregular to spherical stromatoporoids, with wackestone matrix. Top of marginal slope grades into back reef facies. No normal marine bioclasts, much *Stachyoides*. Depth 0–10 m.

10. Bioclastic calcarenite, wackestone–packstone. Normal marine fauna, local onchoids, may alternate with stromatoporoid beds all along the slope and the front of bank. May include displaced back reef *Amphipora*.

Bank interior facies (=back reef facies)

11. Biostromes of *Amphipora* in a pellet mudstone–wackestone matrix with ooids and foraminifera.

12. Pellet calcarenite with ooids and foraminifera but no *Amphipora*.

13. Laminites of fenestral algal mats. Some *Amphipora.* '

Although Lecompte's studies drew upon the enormous wealth of detail that had been gathered by previous workers in the Belgian sections, they also provide an analogy with modern carbonate studies. They gave students of fossil carbonate environments the necessary encouragement at the opportune time in petroleum exploration.

South of the main area of Lecompte's work at the edge of the North Atlantis shelf lies the Rhenohercynian Basin. Within the northern slopes of this basin, however, local *schwellen* or ridges as at Stavelot and Rocroi provided additional sites for reef growth and between 200 and 300 m of carbonate complexes accumulated there. Throughout Middle and Late Devonian times the entire area was prone to rather variable conditions – periodic deposition of red clastic material which blotted out or inhibited stromatoporoid–coral growth on several occasions. These clastic phases mark times of regression and the halting of carbonate deposition on the shelf. Local isolated carbonate mounds survived and were the nuclei from which the carbonate organisms spread out when clear-water conditions returned. With resumed marine transgression shelf biostromes of stromatoporoids spread widely and biohermal growth centres were

again established, depending to a large extent on the nature of the substrate.

An interesting and very widespread 'structure', *Stromatactis*, was described originally (Dupont, 1881) from these rocks. It is in fact a mass of sparry calcite and micritic sediment that is characterised by a smooth basal surface and an embayed and digitated upper surface. The size ranges up to several metres, and commonly *Stromatactis* is associated with *Alveolites*, *Phillipsastraea*, algae and bryozoans. Algal or bacterial activity has been thought to be at least partly responsible for this material, and evidence of an organic origin seems to be increasing.

Distribution of Devonian Carbonate Facies

Devonian shelf or shallow-water carbonate sequences are very widespread in the northern continents but are scarce in South America, Africa south of the Sahara, Arabia, India and Antarctica. Australia, usually regarded as a typical part of the Gondwanaland supercontinent, is atypical in the possession of extensive carbonates. Heckel and Witzke (1979) surveyed the distribution of Devonian carbonates in reference to palaeoclimatic indications (see figure 8.7) using data from Oswald (1968), Vinogradov (1969) and the *Lexique Stratigraphique International*. In their global reconstructions all the major occurrences of the Devonian carbonates appear to conform in position to the tropical belt that is typified by carbonate buildups today, a convincing model but one at odds in places with palaeomagnetic data. The stratigrapher–palaeontologist with an interest in carbonates will probably opt for the Heckel and Witzke model.

North America

The craton was largely covered by carbonate-depositing seas in the Middle and Later parts of the period and evaporite successions occur, as in the Williston and Michigan Basins (see page 115). The Transcontinental Arch, extending during Early Devonian time from northern Canada to north-

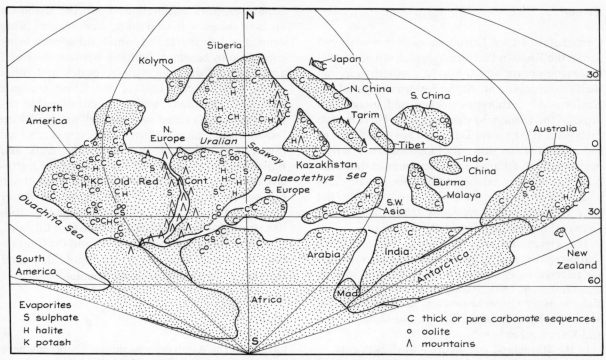

Figure 8.7. Distribution of Devonian carbonates and evaporites: C, thick or pure carbonate sequences; H, halite; K, potash; S, sulphate (after Heckel and Witzke, 1979). The palaeogeographical positioning of the continents is discussed in chapter 10; see also figure 8.12.

west Mexico, was breached so that the Appalachian and Cordilleran seas merged. The carbonates thicken towards the miogeosynclinal Cordilleran and Gulf edges of the continent. Around the margins of the cratonic basins Middle and Upper Devonian typically reefal facies occur. At the edges of the Hay River, Alberta and Williston Basins are vast areas of stromatoporoid buildups and smaller masses are dotted across the basin floors. This began in Givetian time but was affected by the Late Givetian regression that filled the basins with evaporites. Frasnian time was taken up with a series of transgressive incursions from the north which allowed the expansion and proliferation of reefs and carbonate platforms in linear belts and haloes, seemingly controlled by faults or uplifts in the basement. Platform relief was up to 200 m and a variety of reef-associated facies was deposited. The faunal and palaeoecological successions within these facies greatly resemble those described in Belgium by Lecompte. Back reef facies belts are

generally wide and multiple rows of buildups occur. This, together with the absence of oolites, the presence of tidal-flat deposits in some areas and evaporites, suggests quiet hydrographic conditions. Petroleum exploration has revealed the details of many of the individual reefs and banks in the subsurface and the already large literature of this subject continues to grow.

Towards the western edge of the North American craton occur carbonate banks of Late Devonian age that contain rather restricted faunas. They appear to be the source areas for slope deposits fanning out westwards into the miogeosyncline that are locally thick and very coarse. Megabreccias and other flow deposits seem to have moved down slopes perhaps no steeper than 2° and they may have been generated by storm waves or, more probably, by tsunamis or earthquakes. Unlike so many of the carbonate phenomena of the country to the east, these are spectacularly revealed in the ranges of the Canadian Rockies.

Europe

Carbonates of Late Devonian age are widespread across the Russian Platform especially in its central and eastern parts, and along the southern marginal shelf of the continent. At the eastern margin of the platform the carbonates thicken and merge into those of the Uralian Miogeosyncline. There is little carbonate in the Lower Devonian but a thickness of some 200 m of limestones, dolomites and evaporites occurs in the Eifelian. Upper Devonian carbonates, however, are very extensive with subordinate clays and evaporites in the west. They are witness to the Late Devonian transgressions from the Uralian Geosyncline, culminating in the Frasnian.

The Hercynian area of Europe includes both shallow-water and deep-water carbonates, which are referred to in chapter 4 (page 49) and also in chapter 9 (page 160). 'Reefs' are known in Bohemia and the Holy Cross Mountains of Poland, in the Moravian karst area, the Carnic and Dinaric Alps and Karawanken.

In the Rhineland (see figure 8.8) thicker car-

bonate deposits accumulated upon the shallower areas of the geosynclinal floor, these regions being defined and their relatively small subsidence being controlled by basement faulting. Between the reefal areas basinal dark limestones, shales and other clastic rocks occur. East of the Rhine elongate (NE–SW) carbonate bodies of considerable depositional relief existed near the shelf margin (Dorp facies) where almost 1000 m in thickness is achieved (figure 8.9). West of the Rhine the reefs and biostromal limestones crop out; of these the northern examples resemble their Belgian neighbours, but to the south the Eifel area contains flatter mounds on volcanic rises which are capped by atoll-like stromatoporoidal beds. In south-west England, Armorica (Britanny), the Pyrenees and the Cantabrian Mountains of Spain, stromatoporoidal, algal or other organic reefs are conspicuous.

Asia

Largest of the Asian cratonic areas is Siberia where

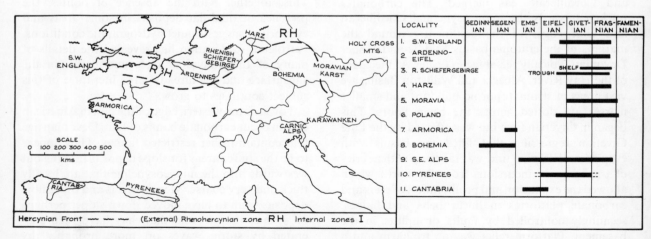

Figure 8.8. Sketch map of Europe showing main areas of Devonian carbonates, and also the position of the Hercynian Front and the boundary between the Rhenohercynian zone and the internal zones of the Hercynian orogeny. The approximate ranges of reefal bank and buildup growth are shown on the right (after Burchette, 1981). Few reefs first appeared in the Rhenohercynian zone before the mid-Eifelian, but in the Internal zones many did so. Few reefs continued to grow after the Late Frasnian here or elsewhere in the world.

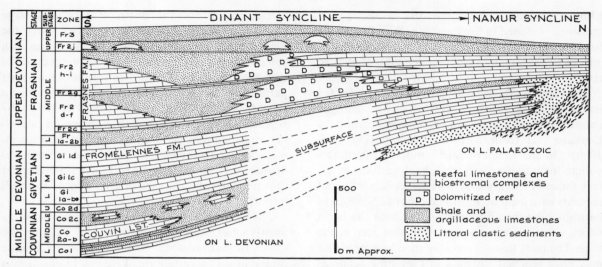

Figure 8.9. Major stratigraphic divisions of the Middle and Upper Devonian in southern Belgium and the Ardenne show conspicuous thinning and overstep northwards. The successive reefs in the Dinant Basin occur in a constant geographical location between broad back-reef lagoons and shale-filled basins. The shale tongues are relatively widespread and are regarded by some authors as marking sudden and repeated rises in sea level, small sedimentary cycles or rhythms being produced (after Tsien, 1977).

thick carbonate successions, hundreds of metres thick, occur around and on the periphery of the craton. Very thick miogeosynclinal accumulations are present in the Taimyr Peninsula and the mid part of the Lena valley and on the margins of the Kazakhstan craton. At the edges of the other cratonic regions of eastern Asia carbonates may reach thicknesses of several thousand metres. Spectacular successions of carbonate occur in South China, and in the other adjacent cratonic interiors and margins. Organic productivity was very high and the rate of buildup was locally correspondingly accelerated.

Gondwanaland

The broad shelf of north-west Africa was largely covered by Devonian shallow seas that left generally rather thin sequences of carbonates. A number of uplifts and basins, all of shallow relief, affected the deposition of both clastics and carbonates. Devonian sedimentation followed the Silurian without break and was locally transgressive to the extent of virtually covering all the ridges and uplifted areas. Local earth movements occurred

between the Lower Devonian and the Eifelian. All manner of Late Devonian carbonate accumulations and reefal structures occur in Algeria, southern Morocco and what used to be Spanish Sahara. The Eifelian to Frasnian bioherms and circular faros in former Spanish Sahara are amongst the most spectacularly exposed, reaching heights of about 100 m and widths of around 1.5 km. They consist of micrite mounds and stromatoporoid–coral facies, and show quaquaversal dips of up to 40° and sharp facies boundaries (Dumestre and Illing, 1968; Hollard, 1968; Legrande, 1968; Elloy, 1972).

Thin carbonates of Devonian age occur along the southern margin of the Himalayas in India and Pakistan but evaporites are unknown there.

Thick Devonian carbonate sequences occur in northern Western Australia, and in eastern Australia where they occur on the continental flank of the Tasman Geosyncline. The evidence suggests that in Middle and Late Devonian time much of Western Australia and the centre of the continent were covered by shallow transgressive seas. The three basins of the north-west reveal Lower Palaeozoic or Precambrian basement covered by transgressive terrestrial and clastic beds. Fully marine conditions only appear to have been established by

Middle Devonian time and the widest transgressions obtained in the Frasnian. Reef formation began at this point and seems to have persisted into the Famennian, which is later than anywhere else.

The Canning Basin (see figure 8.10) is famous for its Givetian and Frasnian reefs which are arranged in linear fashion along a fault system, separating the Lennard Shelf from the Fitzroy Trough (Playford, 1980). On the Lennard Shelf some 500 m of Devonian includes fore reef, reef, back reef and inter-reef facies. Fringing and barrier reefs, atolls and mounds with local depositional dips of 25–30° have been distinguished building out towards the basin which was perhaps hundreds of metres deep at the time. The reefs themselves were sited in waters that may have reached 200 m, and they now occur as exhumed reefal masses for some 350 km along the northern margin of the basin. Originally they may have continued for a further 1000 km. Some of the most spectacular sediments here are the fore reef breccias with clasts up to 100 m across, enormous gravity debris flows that took place down steep slopes. Within the 100–200 m wide reef front are gaps and wide channels and the crest of the reef may be only 60 per cent or less of organic framework. Pinnacle reefs and reef patches occur and the reef front itself forms only a narrow zone at the margin of the platform. The reef front took various forms, being upright, back-stepping retreating or advancing. Stromatoporoids, both massive and irregular, are predominant in the reef structure. Algae were ubiquitous and important builders and binding agents. They maintained a prime role well into the Famennian when the corals and stromatoporoids had declined.

The reason for the death of this very large biogenic structure is not known, but the limestone is succeeded by shales and siltstones with some sandstones. It came after two distinct cycles of reef growth which included interruptions of brief periods of emergence. This rapid transgression in the Middle and Late Frasnian was followed by regression about the Frasnian–Famennian boundary and then renewed transgression in the Famennian. Local tectonics may have had a small influence on the history of the reef but as elsewhere at that time eustatic changes of level appear to be all important.

The basin itself received fine terrigenous deposits, mostly silts, muds and some sand. It includes the Gogo formation which is famous for its unique preservation of vertebrate skeletons in calcareous nodules.

The **Bonaparte** and **Carnarvon Basins** reveal much the same state of affairs, with reefal buildups bordering marginal shelves, and all contemporary with those in the Canning Basin.

In eastern Australia there is a prolific development of carbonates in the miogeosynclinal parts of the Tasman Geosyncline. Here, too, reefal buildups grew at the edges of the coastal platform. Locally the platform edge appears to have been steep, and gravity-flows bearing gigantic carbonate slabs periodically swept into deeper, black-shale environments.

The Devonian of South America contains very few carbonate formations. The cratonic basins are virtually devoid of limestones while only a few thin dark limestones are known within the miogeosynclinal wedges of western Venezuela and western Argentina. The Malvinokaffric area of the Gond-

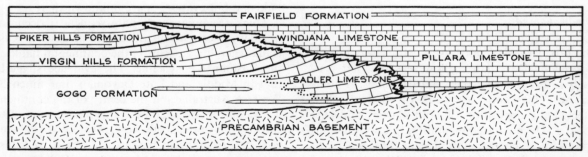

Figure 8.10. The Canning Basin, Western Australia, shows a remarkable succession of Middle to Late Devonian reefs at the edge of the prograding Lennard Shelf (after Playford and Lowry, 1966).

wanaland hemisphere was a cold to cold–temperate zone, largely devoid of calcareous organisms.

Evaporites

Evaporites are so commonly associated with carbonates that it is difficult to separate the two in an account of ancient sedimentary environments. Nevertheless, the conditions that allow the deposi-

tion of evaporites are normally those that kill off the organisms that produce carbonate sediment. We have no single conceptual model for the precipitation of the sulphates and chlorides that make up the vast bulk of the known evaporite sequences. Borates and other complex salts precipitated in lakes feature in a minor way in the stratigraphic column (Till, 1978). They were present in the Orcadian and other lakes of Middle Devonian time, originating when desiccation of these shallow-water

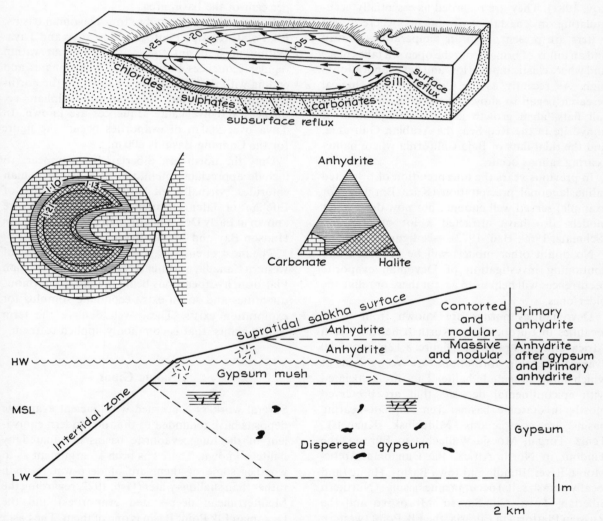

Figure 8.11. (Top) Model of an evaporite basin in which there is a restricted inflow from the open sea. The contours are of water densities and the arrows show the surface influx and subsurface return currents. (Middle) A more idealised model (left) and the compositions of minerals precipitated (triangle on right) (after Briggs, 1958). (Bottom) A model of the littoral and supratidal sabkha (algal) environment (after Kinsman, 1966) relates the distribution of gypsum and anhydrite and the forms they assume.

bodies reached an extreme stage. Recent work in the Dnepropetrovsko-Donetskaya depression in the U.S.S.R. shows Late Devonian evaporites there within a developing rift-valley structure with volcanism in the background. The likelihood of a volcanic influence in the production of many evaporites is strong and a modern instance is the effusion of sodium carbonates from East African volcanoes and their concentration in Lake Natron.

Gypsum-anhydrite, halite and potash salts occur in great quantities in the Devonian system (Zharkov, 1981). They are regarded as essentially accumulating in coastal environments where warm waters are present, high air temperatures prevail and an influx of brines from the open sea takes place and where clastic input is low and net evaporation is high. As recently as the 1960s sedimentological research began to show the close relationships of salt flats, algal growth and the precipitation of anhydrite in the Red Sea, the Arabian Gulf area, and the tidal flats of Baja California where halite-bearing salinas occur.

In previous years the interpretation of restricted saline lagoonal precipitation (Kara Borghaz, for example) served well enough, but now deep-water models also have attracted a lot of attention (Schmalz, 1969; Hsü, 1972) – see figure 8.11.

No doubt other models will be proposed and continuing investigation of Devonian evaporite occurrences will help us to accept these or refine the older ones.

Devonian evaporites are known from many localities in Europe, Asia, North America, North Africa and Australia (see figure 8.12); about 80 evaporite-bearing formations are recognised of which some 26 include halite. They are associated with epicontinental deposits that are preserved mostly in cratonic basins. Ten sulphate-bearing basins are conspicuous – Minusinsk, Kuznetsk, Teniz, Turgai, Moesia-Wallachia in Eurasia, the Tindouf in North Africa, the Canadian Arctic, Moose River, Illinois and Iowa Basins. Halite (and locally potash salts) occur in nine basins – Northern Siberian, Tuva, Chu-Sarysu, Morosovo and the Russian Platform in Eurasia, the Elk Point (western Canadian), Hudson Bay and Michigan Basins in North America and the Canning and Adavale Basins in Australia. (Minor anhydrite–gypsum deposits have also been discovered in Belgium and the western Arctic area of North America.) Of these basins, the Northern Siberian is vast with some 2 500 000 km^2 of evaporites; six are large including the western Canadian Elk Point Basin with hundreds of thousands of square kilometres of salts (see figure 8.13); seven are medium-sized with tens of thousands of square kilometres of salts; and the remainder are smaller. In all of these basins it is general that salt deposition was limited to less than 20 per cent of the total basin-area, except for the Elk Point where halite and sylvite may have covered 35 per cent of the basin area.

Potash salts are limited to four Devonian basins: Elk Point, Russian Platform, Morosovo and Tuva. Within most basins the evaporites occur within regular sequences beginning with carbonates and ending with the most soluble halides; in the northeast Siberian 'giant' some sixteen sulphate sequences and five halite sequences are known. In Tuva over 900 m of evaporites occur; the figure for the Canning Basin is 600 m.

Only the north-east Siberian Basin appears to include appreciable quantities of Early Devonian chlorides, virtually all others contain salts of Eifelian or later Devonian ages. Sulphates are known in Early Devonian formations in Siberia, the Hudson Bay and Moose River Basins.

The most intensively studied basins are those of western Canada, north-east Siberia and the Russian Platform, hydrocarbons being present in enormous quantities and so an extra economic stimulus for exploration exists. These well deserve the term 'saline giants' that is commonly applied to them.

A Saline Giant

Several workers concerned with ancient evaporite deposits have reminded us that no modern equivalent of the huge evaporite basins dominated by halite is known. There has been an argument as to whether some of them are of deep-water origin rather than shallow-water (Hsü, 1972, regarding the Mediterranean deep-seated evaporites) and the Devonian Elk Point Basin is one of them. The basin has been thoroughly investigated not only because oil and sulphur deposits occur within it but also because potash evaporites were discovered there early on. On the one hand, Fuller and Porter (1969)

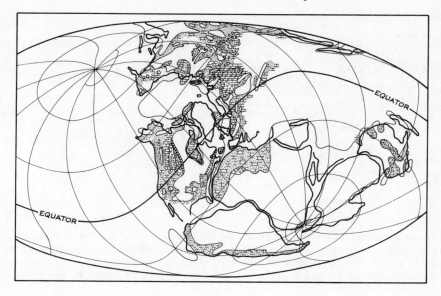

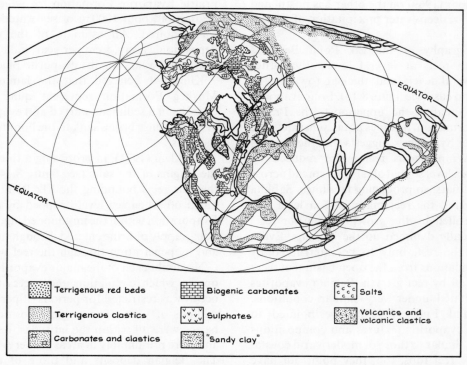

Terrigenous red beds   Biogenic carbonates   c c c c / c c c c Salts

Terrigenous clastics   Sulphates   Volcanics and volcanic clastics

Carbonates and clastics   "Sandy clay"

Figure 8.12. The distribution of Middle and Late Devonian sediments and evaporites as depicted by Zharkov (1981) shows the evaporites generally but not exclusively confined to tropical latitudes (compare with Heckel and Witzke's map, figure 8.7). The distribution of 'sandy clay' in the southern continents largely coincides with what has been regarded as the cold-water Malvinokaffric faunal realm of the Early Devonian.

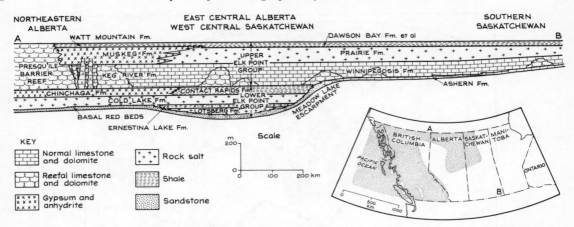

Figure 8.13. A cross-section of the Middle Devonian Elk Point Group in the Prairie Provinces of Canada shows the relationship of the various carbonates and evaporite units and the importance of the Meadow Lake escarpment which was submerged in later Elk Point time as the sea flooded southward. On the location map the stippled part is sea-bounded in mid-Alberta–Saskatchewan by the escarpment.

have advocated a shallow-water origin for the salts while Klingspor (1969) on the other has postulated the origin to be deep-water precipitation of the salt deposits.

The stratigraphy of the Elk Point Basin is complex with reefs at several levels. Evaporites occur at several levels and thicken towards the south-east and east (see figure 8.13). Four evaporite units are mappable; the uppermost, the Prairie Evaporite Formation, includes potassium salts. The shallow-water hypothesis suggests that the lagoons about the Winnipegosis reefs were reduced by lowering of sea level so that algal mats spread across the inter-reef flats, so producing laminites. Sabkha deposits covered the reef margins. Sea level was locally controlled by the growth of barrier reefs which eventually dammed up the lagoons. Evaporation did the rest, only to be interrupted by periodic inundations from the open basin that were soon closed off by reef growth. Belgian evaporites were precipitated under very similar conditions. Shearman and Fuller (1969) have pointed to nodules and evaporitic textures and compositions that are very similar to those of modern arid coastal sabkhas. The reef tops, too, they point out have been penecontemporaneously weathered. As evaporation reached extremes so potash salts were precipitated, a gentle supply of sea water entering over the Keg River–Presq'ile barrier reef.

Almost 200 m thickness of salts is present in the Prairie Evaporite Formation for which the evaporation of about 1000 m of sea water in the basin was required. An estimate of the rate of salt deposition as, say, 5 cm per year, indicates a life of some 4000 years for the evaporite phase.

Klingspor believed that the laminates in the Muskeg Formation represent quite deep-water deposits in a sediment-starved and evaporitic basin in which dense brines sank to the floor to precipitate the salts.

Maiklem (1971), in proposing a third version of the origins of the salts (see figure 8.14), suggested that the reefs bounding the Elk Point Basin were killed off when sea level dropped by about 30 m. Evaporation within the impounded lagoon raced on but a supply of brine entered through tidal channels and by penetration through the reef walls.

No doubt each of the major evaporite basins was one in which circulation and access to the open ocean was restricted for periods of perhaps tens of millions of years. The original positions of these basins were all within the latitudes well below 60° and were presumably in areas of (very) low rainfall. The Belgian deposits and those of the Russian Platform are nevertheless associated not only with carbonates but also with terrigenous clastics and some carbonaceous strata; a humid warm climate may be indicated. An explanation of expansion of

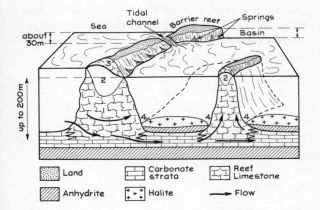

Figure 8.14. Maiklem's (1971) model for the origin of evaporites in the Elk Point Basin postulates barred lagoons lowered by evaporation and fed by percolating sea water and tidal channels. Normal sea water enters the reefal limestone (1) beneath a fresh-water lens (2). The top of the reefal limestone maintains a vadose zone but normal sea-water flow at (4) causes solution of the precipitated salt layer.

areas in which excessive evaporation took place between Middle and Late Devonian times has exercised many workers. The Russians have set much store on the fact that evaporites and terrigenous red beds occur mainly in four regions (northern North America, northern Europe, Angaraland–Altai Sayan and Australia) which they regard as arid, although not all geologists see these areas as equally so. Even the most recent Soviet discussion of the place of evaporites in Devonian palaeogeography (Zharkov, 1981) makes little use of palaeomagnetic data to establish the position of the continents. Nevertheless, the conclusion reached is that Devonian climatic arid zones and humid zones did not form a continuous belt or belts but were controlled by the arrangement of oceans, seas, coastlines and land relief as well as patterns of sea and atmospheric circulation.

The study of Devonian carbonate and evaporite deposits has prompted many interesting depositional models, provoked argument between palaeontologists and sedimentologists and broadly influenced most palaeogeographic world maps. The essentially organic origin and control of carbonate reefs has been emphasised by authors such as H. H. Tsien; others have addressed themselves to a more sedimentological approach. Evaporite precipitation models are equally varied: both may provide valuable information about sea level and strandline movement. Increasingly detailed studies of both carbonates and reefs now reveal depositional histories in which these movements are more numerous and widespread than had been suspected.

In the literature about carbonate and evaporite sequences the subject of cycles within them is commonly given much space and the causes of the cyclicity have been ascribed to tectonic (epeiric), eustatic and climatic factors. Progressive desiccation may be interrupted by periodic influxes of fresh water from the adjacent land mass or by rise of basin water level. In most instances it is difficult to determine the cause of the stratigraphic repetition of evaporite deposits.

Patterns of deposition within the Middle and Late Devonian, however, indicate a general tendency towards widespread transgression almost universally until the end of the Frasnian. The sudden regression occurring then terminated reef building and carbonate production with a puzzling rapidity. This is an event to which we shall return later.

Carbonate and evaporite deposition on the scales indicated in the Devonian system must have had effects upon the geochemical cycles of calcium, carbon and sulphur, and the alkali metals, sodium and potassium. A lowering of world ocean salinities as a consequence of rapid evaporite deposition may have occurred and could have had climatic as well as biological effects. The precipitation of major evaporite bodies, as Garrels and Perry (1974) have pointed out, may necessitate huge transfers within the sulphur and carbon reservoirs. Precipitation of calcium sulphate at rates higher than normal (steady-state) transfers Ca^{2+} from the carbonate to the evaporite reservoir and back, with a net gain of CO_2 in the atmosphere. In the steady-state world this gain would soon be offset by increase in burial of organic carbon, by increased dissolution of carbonates, or both. Should evaporite deposition be widespread and rapid this balance may not be achieved and, in the short term, excess CO_2 may remain in the atmosphere. A climatic warming may have been followed by a quick response in the plant kingdom—witness the Middle and Late Devonian forests and coal deposits. The vigour of that response may have had repercussions in the evolution of the vertebrates and especially those that were to become terrestrial and air-breathing. As CO_2 levels were reduced and the O_2 content of the atmosphere rose, further vertebrate response may have followed. Meanwhile atmospheric composition would perhaps also have had effects upon the marine floral and invertebrate populations.

9

Deep-Water Suites

Under this heading are considered rocks that are generated in deeper waters beyond the continental shelf, commonly distant from active sources of terrigenous materials, and in the oceanic basins. Pelagic and bathyal accumulations, the deposits of oceanic floors and the continental slopes, are of this kind. The basins themselves may be extremely large, underlain by oceanic or intermediate crust; they may be broadly uniform throughout or interrupted by island arcs or deeper oceanic ridges and volcanic structures. While some were of great duration others have been short-lived, quickly filled by sediment or closed by diastrophism. All are of importance in earth history.

Although both terrigenous clastic and carbonate deposition is highly variable across the continental shelf, however over the continental slopes and across the oceanic floors pelagic sedimentation becomes relatively more important. There, beyond the 100 fathom or 200 m contour the rate of sedimentation is normally very slow. Two or three centimetres per 1000 years is perhaps normal, fine clastic sediment being predominant and pelagic carbonate rare. Pelagic organisms contribute an incessant, mostly microscopic, rain of organic debris but there are comparatively few creatures in the benthos. Normally, deep basinal environments experience little diurnal or seasonal disturbance, and they are cold, aphotic and euxinic. They may nevertheless experience periodic interruptions of a violent kind when submarine sliding, slumping, mass flows and turbidity currents occur, deposits of more than a metre thick bringing the rate of deposition up to many metres per 1000 years. Modern analogues for these flysch accumulations are now well known.

Basinal associations may include large volumes of interbedded volcanic material, either pyroclastic or in the form of basaltic lava flows. In areas of tectonic activity, such as island arcs, volcanic activity is endemic, diverse, and frequently occurs close to rapidly subsiding areas of ocean floor (that is, basins of a subduction trench type). A feature of such areas of relative instability is the frequency and volume of material that is shifted by mass movement of one kind or another into the deeper waters from the shallow margins of the basins or from positive regions, uplifts, volcanic highs etc.

These phenomena are recorded from innumerable instances in the stratigraphic record, and largely from those thick prisms of strata referred to as geosynclines. As noted in chapter 3, although 'geosyncline' has become part of plate-tectonic theory its ghost has not disappeared. Seemingly abnormally thick accumulations of sediment at continental margins are still referred to by this term. The conditions under which they develop may be quiet, as on Atlantic-type continental margins, or they may be highly changeable with varying degrees of tectonic activity and volcanism (Pacific-type). Many of these accumulations have suffered extreme deformation in subduction zones, orogeny and continental collision. With an end to subduction, isostatic uplift and plutonism may continue to affect these materials; they become 'stabilised', accreted to the edge of a continental mass.

In the Phanerozoic tectonic belts the great prisms of sediment that accumulated under 'geosynclinal' conditions are revealed as deformed masses with major dislocations, ophiolites, plutons etc., obscuring the nature of the original basin in which the material was deposited. Some of these belts persisted throughout much of the eon and Devonian sediments were only part of the great sedimentary input. This has been the case especially in the belts that surround the Pacific and in the Mediterranean–Himalayan region.

Deep-water Diagnostics

Oceanic and deep-water basins in the past may be expected to have received not only the finest terrigenous material from the continental areas but also pelagic sediments, resedimented coarser clastic debris and fine volcanic and meteoric materials (see figure 9.1). Black anoxic mud lithologies, typical of the 'sediment-starved' parts of basins, may not vary much between relatively shallow and deeper environments. The problem of black-shale accumulation has been discussed above (page 121).

The true pelagic sediments may not be easy to recognise, being perhaps relatively inconspicuous or only minor components of the basinal accumulations. They commonly contain more than 30 per cent of biogenic pelagic material. (Modern oceanic oozes are examples, while chalk limestone and cherts both represent ancient pelagic but not necessarily deep-water accumulations. Chemical deposits in the basin environment include manganese nodules and phosphorites.)

The earliest recognisable true pelagic sediments in Europe may be the Silurian *Orthocerenkalk* of the Carnic Alps. Siluro–Devonian orthocerid-rich, thin, phosphatic limestone in southern Morocco appears to be pelagic in origin, but late in the Devonian period pelagic carbonates became widespread in Europe and North Africa. Not all originated in deep water so much as in areas that were

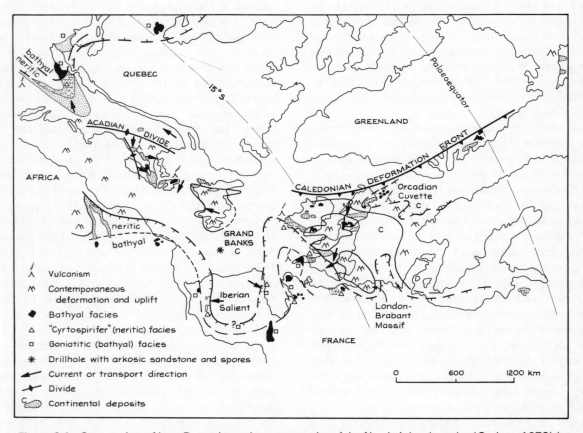

Figure 9.1. One version of Late Devonian palaeogeography of the North Atlantic realm (Greiner, 1978) in which the deeper water ('bathyal') facies are indicated. Note the prominence of an 'Iberian Salient' shown as perhaps part of a North Atlantis continental mass. Other authors have regarded Spain as being, together with Italy and southern Europe, parts of the Gondwanaland continental mass.

cut off from terrigenous sources by deep basins, and that lacked benthonic faunas.

Fossils are locally abundant in Devonian basinal rocks, and in Europe have prompted such terms as *Tentaculitenschiefer, Styliolinenschiefer* and *Cypridinenschiefer*. Plankton must have been extremely abundant in many parts of the oceans since the benthos were locally confined. There were rich phytoplanktonic communities. Epiplanktonic animals included possibly byssate bivalves attached to sea weeds (for example, '*Posidonia*' *venusta*, *Buchiola*), inarticulate brachiopods in large number, and some small gastropods. Among the nekton ammonoids, nautiloids, 'tentaculitids' and ostracods and the conodont animals were foremost. It has been suggested that the dwarfed ammonoids so common in deep-water black shales are immature forms which fell into an anoxic environment from the breeding sites above. Graptolites, present in the Early Devonian plankton, were nowhere as abundant as they had been in earlier Palaeozoic times.

One of the most spectacular types of marine sedimentation must be that achieved by high-density turbidity currents. In recent years a great deal of information concerning modern turbidity current activity, and the sedimentary bodies they produce, has come from studies on the continental slopes and oceanic plains of the Atlantic and Indian Oceans. Submarine canyons and submarine (abyssal) fans have been surveyed and studied. Their size may be enormous, as in the case of that, many thousands of metres thick, off the Ganges delta in the Bay of Bengal. In the sedimentary record the presence of turbidity current deposits has been known for many years and even from the earliest days of geology the thick successions of flysch have been a source of interest. The fan-like overall shape of several such accumulations has been confirmed, though it should not be assumed that all greywacke bodies have this form.

A well-documented Devonian example is from western Argentina where Gonzalez-Bonorino and Middleton (1976) report the Punta Negra Formation to consist of a monotonous greywacke and shale sequence (see figure 9.2). It thickens westwards to reach a maximum of 1700 m, and was carried by palaeocurrents from the south and southeast. The greywackes form beds that are 0.5–1 m thick, arranged as several packets or groups separated vertically by several metres of shale. They contain 40 per cent rock fragments and 20 per cent matrix, seemingly derived from a terrain of Precambrian metamorphic rocks. Sand was transported across the Punta Negra fan by high-density turbidity currents and the fan was developed between two north–south-trending submarine rises. There seems to have been no single main feeder channel.

Turbidity current deposition may have given rise to other lithological types. Resedimented crinoidal debris sheets are known in Cornwall (Tucker, 1969). In Galicia, north-west Spain, calcareous material comprises thick strata in the greenschist facies belt around Cabo Ortegal, and had been regarded as of Precambrian age (Meer-Mohr, 1975). The identification of coral fragments, crinoidal debris and foraminifera now provides a Siluro–Devonian date for this material that was deposited locally in deep submarine troughs.

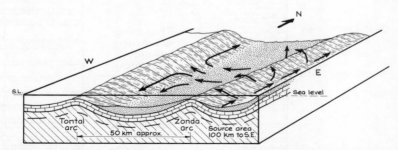

Figure 9.2. The Punta Negra basin received much of its turbidite infilling from flows that travelled north to the east of the Zonda arc and then crossed the arc to become 'ponded' by the Tontal arc (after Gonzalez-Bonorino and Middleton, 1976).

Volcanic materials may be subjected to mass flow and resedimentation in deeper-water environments. The Merrions Tuff, for example, is an Early Devonian volcaniclastic unit that occurs within a flysch-like sequence in New South Wales, Australia. It consists of predominantly grain-supported feldspathic crystal aggregates or arenites organised into very thick groups that are of widespread extent. Graded bedding is only to be seen in beds that are less than 2 m thick but the general characteristics of the unit suggest large-scale submarine mass flows and turbidity action in a deep interarc basin within the Tasman Geosynclinal area (Cas, 1979; Cas *et al.*, 1981, and see page 170 below).

Submarine debris flow deposits, slumped beds and turbidities are associated with the palaeoslopes of many ancient continental margins and oceanic rises (see figure 9.3). Flysch, long considered to be the characteristic sedimentary suite of lithologies of deeper water, is locally accompanied by other types of accumulation where gradients, sediment supply and local tectonics are suitable. Debris flows on a grand scale are present in many geosynclinal sequences throughout the stratigraphic column. Devonian examples are known in Western Europe, Morocco, eastern Australia, China and elsewhere.

Padgett *et al.* (1977) describe a series of local Late Devonian and Early Carboniferous debris flows near Rabat in western Morocco. These exhibit many of the characteristic features of such flows, and they lie adjacent to fault scarps that were active and probably generated turbidity deposits. The mechanism that triggered off the collapse of large masses of material in blocks was probably sudden fault movement during the so-called 'Hercynian Orogeny' in this area. The collapsed debris came to rest on the flanks of the upthrown sites from which it originated.

A more problematic instance of mass movement into an active basin is the Rinconada Formation of western Argentina (Amos, 1954). This Early Devonian assemblage of coarse conglomerates, sand-

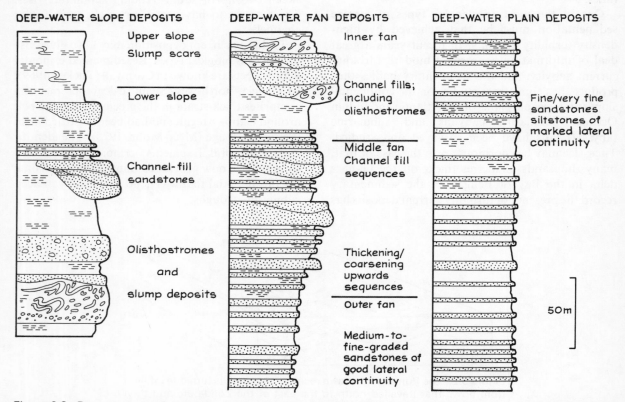

Figure 9.3. Deep-water environments may be characterised by particular sequences of clastic sediments. Slope and fan environments may contain slumped and collapsed deposits (after Mutti and Ricci-Lucchi, 1972).

stones and shales also includes innumerable 'exotic' masses of Ordovician limestones, in size anything up to slabs 2 km long and nearly 200 m thick. Amos suggested that the Rinconada conglomerates were fluvial debris that was dumped rapidly into a geosynclinal foredeep while the Ordovician material gravitationally collapsed from the tectonically active edge of the basin. Today the term 'olistostrome' might be applied to this formation, but the combination of local gravity collapse debris and great fluvial conglomerate input is perfectly possible.

Devonian Basins

Devonian basinal deposits of oceanic origin, that is open water if not actually bathyal accumulations, identified by their distinctive lithologies, tectonic settings and condition and by their pelagic or nektonic faunas, occur in many parts of the world. They are known in greatest detail in Europe. Most of them are situated towards the top of sedimentary masses that had been accumulating since well before the beginning of the period. They can be listed broadly as follows:

Central and Western Europe (Rheic Ocean deposits) (see page 168)
Appalachian (Foibic Ocean)
Cordilleran
Franklinian
Urals (Pleionic)
Alta-Sayan and Angara
Altai–North China (Junggar-Hingan region)–north of the Inshan-Tianshan Mountains
South-west China–Himalaya–West Yunnan and West Sichuan; North Xizang (Tibet)
Mediterranean–Caucasus (Rheic Ocean)
South America, South Africa, Antarctica and Australasia ('Samfrau Geosyncline')
West Pacific

Central and Western Europe

A summary account of the Devonian basinal facies of central Europe by Krebs (1979) is particularly interesting because it was in this area that Devonian biostratigraphy was largely established and because it concerns a (geosynclinal) region with a wide range of facies that were all thought to have accumulated below the 200 m line and for the most part well below the level of the adjacent continental shelves. Pelagic sediments also occur as the Griotte limestones and as flysch in the Pyrenees (Tucker, 1974) and Montagne Noire areas and in Menorca, the Rif and Tuscany, in south-eastern Europe and in south-west England (see figure 9.1).

In central Europe these facies occur in areas that lie between the continental shelf of the North Atlantis (part of Laurussia) land mass (the 'external shelf' of Krebs, 1974) to the north, and the crystalline belts and/or volcanic arcs to the south (the 'internal shelf'). Throughout its development this geosynclinal basin received influxes of clastic material, largely sandstone and limestone as thick shallow-water clastic sediments to the north. There followed a brief regressive phase with red beds and the Ems Quartzite being deposited, but with the Wissenbach Shale sequence the spread of deeper water was extensive and in the Late Devonian the Ostracod Shale facies became very widespread. Black-shale horizons can be traced from the basin on to the shelf areas and may reflect change of sea level with a corresponding rise in the level of the thermocline.

Over the submarine rises within the basin thin, partly condensed, pelagic limestone facies accumulated. They are usually known as '*Cephalopodenkalk*' because of the numbers of goniatites and orthoceratid fossils within them. Generally they consist of well-bedded, locally nodular grey, black or red limestones with thin shale interbeds. Laterally they grade into the basinal shales and siltstones.

Within the more mobile parts of the region the basinal sedimentation was of shale, with minor silt and sands. Dark colours prevail and the fossils are graptolites, conodonts, fish, thin-walled bivalves, tentaculitids and styliolinids with a few smooth brachiopods and trilobites. The coarser clastic rocks are turbidites from a shelf margin or from local sea mounts or atolls. In the Rhineland some 1500 m of Middle Devonian black shale correspond to only about 150 m in the Saxothuringian region.

Tucker (1973), in his study of the sedimentology

and diagenesis of the *Cephalopodenkalk* and associated sediments that occur on the schwellen of the Rhenohercynian Geosyncline, believes that these condensed pelagic limestones were formed at depths from about 50 m to a few hundred metres. They pass laterally into nodular limestones with shales, commonly slumped and reworked. This slope facies passes down the gradient into silty shales and local turbidites which were deposited in basins about 1000 m deep. Similar pelagic limestones are known in the Halstadt facies of the Alpine Jurassic, and similarly are considered to be ancient analogues of present-day sea-mount limestones. Early cementation seems to take place in all of these as indicated by hard ground surfaces, sheet-cracks and neptunian dykes as much as 0.5 m deep.

Coarser sediments present are calcareous turbidites, greywackes or mature sandstones. Styliolinid-rich pelagic limestones, studied by Tucker and Kendall (1973), have been likened to modern pteropod oozes, but complex diagenesis has converted the original carbonate to microspar. (A single thin schwelle-like occurrence of nodular limestones with shales and a complete suite of the

Famennian zonal ammonoids has been found at Chudleigh in South Devon, to the south of which are typical basinal ostracode shales and pillow lavas (House, 1963).) Beneath the limestones the 'basement' may be a submerged reef or neritic carbonate feature, a volcanic ridge or a geanticlinal structure; all have a north-eastern trend parallel to the general axis of the geosyncline. Between the rises, which are only a few kilometres across, the basins are several times as wide. Within the limestone is a fauna of ammonoids, conodonts, turbidites from the north, and pulses of conglomerates, flysch greywackes, olistostromes and volcanic materials from the south. Palaeocurrent directions are from both margins of the basin and also along its axis, and detrital material was also locally spread about 'scattered islands, sea mounts and atoll-like reefs developed within the basin' (Krebs, 1979). The submarine rises within the basin have long been known as '*Schwellen*' (see figure 9.4) and the lower basinal parts as '*Becken*'.

Various tectonic settings have been postulated for the development of this facies, among them being a trench–abyssal plain environment devel-

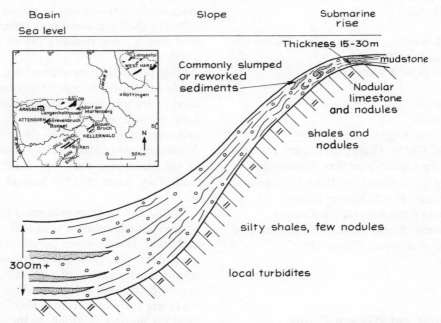

Figure 9.4. Devonian 'Schwellen' (pelagic) carbonates of the Rhenishe Schiefergebirge are indicated (black) on the inset map. Their facies and lithological relationships relate to deposition in shallow water with lateral resedimentation into deeper areas.

oped on oceanic crust, a marginal oceanic basin formed by back arc spreading, and a floor of metamorphosed continental crust that is subsiding because of continental rifting. During the Devonian period the areas of maximum sedimentation within this region shifted periodically northwards unlike those of the Saxothuringian area and of Armorica.

The onset of deep-water conditions is represented in the famous (Lower Devonian) Hunsrück Shale which interfingers with the dacryoconarids, ostracods and thin-shelled posidonid bivalves. Microsparitic and biomicritic limestones are the most common types, cut by abundant shaly streaks (Flasers) up to a few centimetres thick.

In the Upper Devonian the pelagic limestones range between 15 m and 30 m in thickness whereas the basin sediments are more than 300 m thick. They pass laterally into nodular limestones and shales known as *Cypridinenschiefer* on account of abundant ostracods. In origin the *Cephalopodenkalk* appears to be rather coarse pelagic sediment and its condensed nature is the result of current activity. The entire Upper Devonian is represented

by a mere 10–30 m, one-tenth or less of the thickness accumulating in the adjacent deeper waters. The nodular limestones may have been produced in a variety of ways—by unmixing during slumping, compaction and spreading, subsolution of limestone, local limestone precipitation around bacterially decomposing matter and by diagenesis.

The local Becken or basin sediments are conspicuous red to black or grey silty shales; some are calcareous, others are carbonaceous but the most conspicuous and common are thinly laminated shales. The laminae fine upward into clay–shale and have sharp bases, being the deposits of low-density turbidity currents. Clearly no major clastic influx interrupted the deposition of these beds: they are thought to have accumulated in 'starved basins' that were marginal to the site of turbidite deposition. Pelagic body fossils are very rare but trace fossils suggest that the sediment was oxygenated enough for some animal life.

Figure 9.5 shows a cross-section of Devonian rocks in the German outcrop.

The southern limit to the European basinal

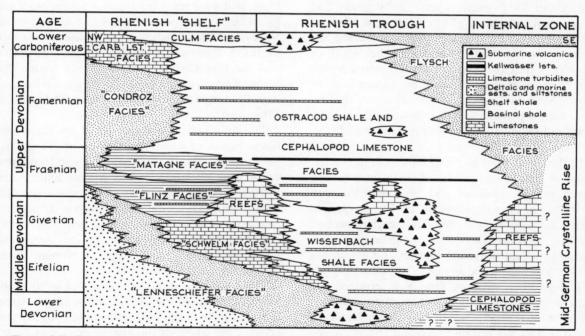

Figure 9.5. The great interest that the German outcrops of Devonian rocks hold is well illustrated by this cross-section. It shows facies distribution from the epicontinental environment into the deeper water facies of the Rhenish Schiefergebirge and the south-east (after Krebs, 1971). The 'Condroz facies' and 'Lenneschiefer facies' are fossiliferous sandstones; the 'Matagne facies' and the 'Flinz facies' are shales: the 'Schwelm facies' include biostromal and bituminous limestones and interbedded shales.

deposits (and hence of the Rhenohercynian 'geosyncline' or ocean) is uncertain (see page 180). Devonian strata in North Africa are of epicontinental shallow-water facies, and from here the shoreline passed east into southern Turkey.

Between the Silurian and the Carboniferous thick, deep-water carbonates and shales, said to have been deposited in water 200–4000 m deep (Bandel, 1974), in the Carnic Alps show that periods of deposition alternated with those when it was negligible and when bioturbation, ferruginous crust growth and resedimentation took place. Eastwards from Bohemia and Austria shoreline facies extend into the Balkans and to the Caucasus.

The Appalachian Basin

Stretching from Newfoundland and maritime Canada to Alabama, the Appalachian region includes several structurally distinct provinces — the north-eastern area of upland New England and eastern Canada, the Piedmont, Blue Ridge and the Valley and Ridge sections and the Allegheny and Cumberland plateaux. North from New York to Newfoundland the basin was closed during the Caledonian and Acadian orogenies, but south of New York the closure came later. The entire region has been intensively studied by generations of North American geologists and was one of the first for which a plate-tectonic explanation for its tectonic and sedimentational history was offered (Dewey and Bird, 1970).

In the north (that is, Newfoundland) the earlier orogenic phases had produced small local basins, which rapidly subsided and became filled with clastic red beds and volcanic rocks. Eugeosynclinal facies are also present here. In nearby Quebec, however, Lower Devonian sedimentation was miogeosynclinal with thick clastics and limestones (2000 m) in Gaspé and shales to the west.

The Acadian orogeny largely determined the pattern of Middle and Late Devonian sedimentation in the northern part of the Appalachian belt; south of New York State its tectonic effects are not so apparent. However, the elevation of the Acadian mountain upland provided an immediate source of sediments which flooded westwards in the New York–Pennsylvania area to give the great clastic wedge of the Catskill Delta. South of this rapidly accumulating pile and to its west the coarse clastic rocks give way to shales and limestones that are indicative of marine conditions and deeper water. The reduction of thickness and the accompanying facies change have become classics of their kind (see, for example, the masterly summary by Rickard, 1975).

The miogeosynclinal belt of the Catskill Basin and the Appalachian provinces to the south of it interposed between an uplifted Appalachian land mass to the east and the edge of the American Craton to the west. During the remainder of the Devonian period this basin received continuous supplies of fine marine clastic sediments and several inputs of carbonates. All thin westwards on to the cratonic interior and southwards towards Alabama and beyond. Black shales occupy an increasing proportion of the succession in those directions. The regular alternation of black shale units with others has long been known as a mark of persistent cyclicity within this basin throughout the period. Many of the carbonates are highly fossiliferous; the black shales are regarded as a rich source of oil and gas. Thus the miogeosynclinal basin in the present Valley and Ridge province very efficiently trapped the influx of terrigenous sediment from Appalachia and the thin Upper Devonian black shales that spread westward on to the eastern part of the craton have been regarded as typical of a 'starved basin' in consequence.

The only known volcanic rocks are three bentonite bands, of which the middle one (Tioga) has been recognised over a very wide area and is of great value for correlation in the Appalachian Basin. Further to the south, rocks of the now separately outcropping Ouachita Basin suggest a possible continuation of the Appalachian belt. The Lower Devonian there is represented by limestones and novaculite; Middle and Upper Devonian strata have not been recognised there.

The eventual tectonic rumpling and uplift of the miogeosyncline of the central and southern Appalachians came in the Late Palaeozoic orogeny that affected the whole of the eastern margin of North America.

The Cordilleran Geosyncline

This very long-lived and major geological feature is part of the history of the great circum-Pacific orogenic belt. An eastern, miogeosynclinal, part received a thick column of clastic and carbonate sediments between Late Precambrian and Late Palaeozoic times but underwent a *décollement* type of deformation rather than metamorphism and igneous activity. The western part of the Cordilleran orogen was the site of eugeosynclinal sedimentation which was at times disrupted by narrow geoanticlines, themselves affected by granitic intrusions and intense regional metamorphism and deformation. The troughs acquired volcaniclastic and clastic sediments and contemporaneous faulting and folding took place. Present-day mountain chains forming part of this great complex and containing important Devonian deposits are the Alaskan Coastal Ranges, the Canadian Rockies, and the Central and Southern Rockies (see figure 9.6).

Miogeosynclinal sedimentation in Devonian time was marked by an expansion and transgression of the sea eastwards. Thin basal clastic beds are overlain by thick carbonates and some evaporites in British Columbia. Thicknesses of more than 3000 m of Lower Devonian occur in the Mackenzie Mountains.

Eugeosynclinal rocks of Devonian age are widespread and include thick black shales, limestone lenses, thick cherts and greywackes, some conglomeratic spreads and enormously thick andesitic and basaltic volcanics. In both Alaska and Nevada unusually thick cherts (2000 m) have been noted which resemble the novaculite of the Ouachita geosyncline.

Various parts of the geosynclinal belt suffered uplift and erosion during Devonian times, some of the eroded material being resedimented relatively locally. For example, evidence from several parts of central Alaska indicates the presence of the Devonian outer continental shelf, continental slope and the ocean basin draped with turbidites and shales, and thin bedded chert in the basin. A deep-sea turbidite fan of very large proportions is suggested; its inception may have been early in the period; it lasted until late. A dismembered ophiolite complex in the Livengood area may be a chunk of Devonian oceanic floor that was carried up on to the continental edge.

In the northern Yukon orogenic spasms, which seem to relate to the Ellesmerian Orogeny of the Franklinian regions, occurred in Late Devonian and Early Carboniferous times; they produced some metamorphism and folds, and caused the intrusion of small Givetian and Frasnian-aged plutons.

In the latitude of Nevada other compressional effects were being produced by subduction; the orogenic phase known as the Antler was responsible for the Roberts Mountain and other thrusts which

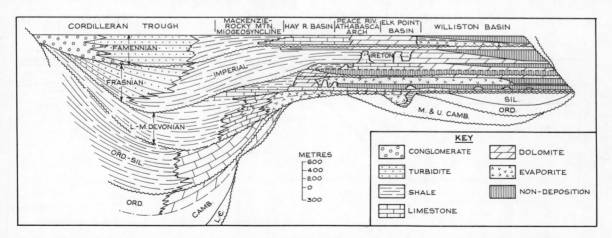

Figure 9.6. The Devonian basins of western Canada range from the deep Cordilleran trough to the shallow epicontinental evaporite basin of the Prairie Provinces (Williston Basin). This cross-section shows the principal formations found and the widespread unconformities that define the Kaskaskia Sequence.

pushed deeper-water Devonian facies eastwards over those on the shallow cratonic shelf edge (see figure 9.7). Extensive outcrops of spilites, keratophyres and quartz keratophyres occur amongst rocks of Devonian age in California, Nevada, Oregon and Washington, but plutons of this age are not known.

The Franklinian Geosyncline

The northern islands of the Canadian Arctic Archipelago are occupied largely by the Sverdrup Basin, a regional depression holding rocks of Carboniferous to Cretaceous age, but from beneath this structure emerge parts of the Franklinian Geosyncline (Thorsteinsson and Tozer, 1968; Trettin, 1972; Christie, 1979; Kerr, 1982). They reach from Melville Island in the west some 2100 km to eastern Ellesmere Island and northern Greenland. On the southern side the strata are contiguous with the Palaeozoic platform deposits of the stable Arctic Lowland Platform. The geosyncline includes a eugeosynclinal belt to the north of a broader miogeosynclinal belt, and accumulation lasted from Late Precambrian to Late Devonian times. A thickness of 5000 m or more of Devonian is present.

Extending north from the Arctic Lowlands towards the centre of the Sverdrup Basin is the Boothia Uplift which has raised the Precambrian basement about 8000 m. The uplift appears to have been active on a number of occasions throughout Devonian time. The sediments throughout were derived from the north-west, where lay the Pearya Mountains or Geanticline, an intermittently rising orogenic arc. The axial area of the Franklinian Geosyncline between northern Greenland and north-western Melville Island shows that the deep-water geosynclinal sediments overlie shallow-water deposits rather than oceanic basalts and hence the trough developed by subsidence of the continental crust rather than by a sea-floor spreading mechanism.

The miogeosynclinal rocks bordering the Arctic Lowlands comprise thick carbonates, quartzose sandstones and shales. Facies relationships are complex with Silurian and Devonian carbonates grading laterally into argillaceous rocks locally with cherts. In the vicinity of the Boothia Uplift occur the conglomeratic red beds of the Peel Sound Formation that were described on page 106. General facies relationships and thicknesses are shown in figure 9.8. Middle and Upper Devonian strata are widespread in the miogeosyncline with carbonates,

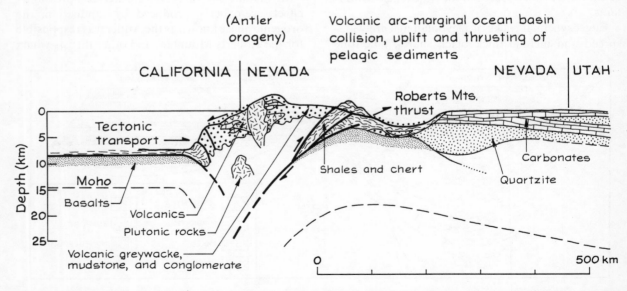

Figure 9.7. The Devonian Antler Orogeny on the western seaboard of North America exemplifies an orogenic continental margin of the Pacific type where a volcanic arc is pushed against the edge of the craton. Greywacke sedimentation is thereby replaced by molasse shed from the new uplands (after Churkin, 1974).

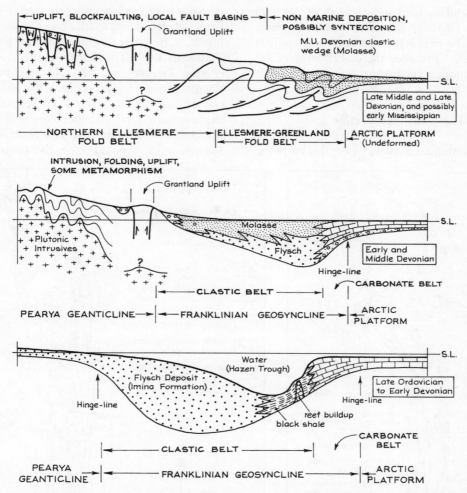

Figure 9.8. During Devonian time the long-lived Franklinian Geosyncline became largely flysch-filled and then, with the rise of the Pearya Geanticline mountains, flooded by molasse that eventually spread on to the Arctic Platform (after Kerr, 1982).

evaporites and shales dominating in the lower part and a great clastic wedge developing in the upper (see page 105).

On the bathymetric axis of the geosyncline from Early Ordovician to Early Devonian time was the Hazen Trough in north central Ellesmere Island. The Devonian eugeosynclinal rocks there are thick greywackes and other immature sandstones, shales, phyllites and cherts and thick volcanics, and deep-water sedimentary rocks extend subsurface as far west as Banks Island. Some 3500 m of continental Lower to Middle Devonian red beds are followed by 3500 m of the Svartevaeg conglomerates, turbidites and volcanics in Axel Heiberg Island. The

coarser clastic sedimentary rocks are submarine slump deposits which include Silurian carbonate clasts. They suggest sudden downwarping of the basin.

All of these rocks were affected by the Elles-merian orogeny at the end of the Devonian period, the Parry Islands and Central Ellesmere Fold Belts being produced. Earlier movements in the eugeo-synclinal area extend back at least to pre-Devonian time. Uplift of the deformed and metamorphosed eugeosynclinal area was probably instrumental in feeding detritus to the great clastic wedge of the miogeosynclinal region in Middle–Upper Devonian times. Small granite plutons in northern

Ellesmere Island include some dated as 360 ± 25 million years (that is, Late Devonian). It seems entirely probable that orogenic events and uplift in the eugeosyncline preceded the deformation of the rocks in the miogeosynclinal area.

The Ural Geosyncline

The Ural Mountains and Novaya Zemlya mark the site of a long-lived mobile belt, over 3000 km long from the arctic coast south almost to the Sea of Aral (figure 9.9). Separating the Russian and Siberian Platforms, the belt has been recognised as the site of (Permo–Triassic) collision of these two subcontinents. Morel and Irving (1978, pp. 551–552) point out that on the face value of palaeomagnetic data the northern Ural trough may have been quite narrow, but that an anti-clockwise rotation of Siberia, equally possible on the geophysical evi-

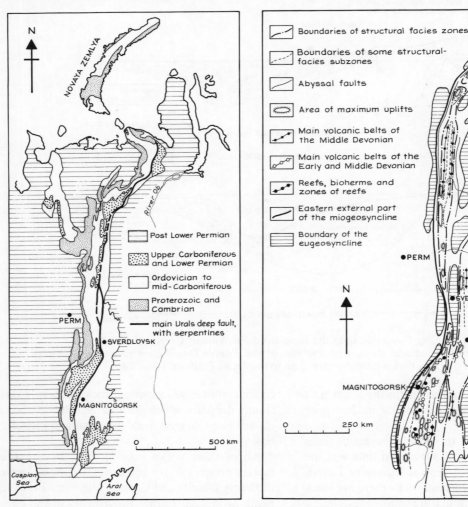

Figure 9.9. Left: The main geological features of the Ural mountains belt in which the so-called miogeosynclinal zone or western flank of the Urals and the so-called eugeosynclinal zone or eastern flank are separated by the Urals deep fault (suture?) zone (after Read and Watson, 1975). Right: Tectonic zones of the Uralian eugeosyncline in the Middle Devonian. Six such zones are distinguished by their structures and facies and each developed as a separate entity. Subsidence and vulcanicity are linked features but the total thickness of Devonian sediments scarcely exceeds 2.5 km. Areas of minimal subsidence are characterised by biohermal formations (after Breivel et al., 1968).

dence, would suggest that the Urals represent a large ocean (Pleionic Ocean) east of Baltica. The truth of the matter has yet to be determined. Within this belt both miogeosynclinal and eugeosynclinal assemblages are present. Structurally it is highly complex and has the Urals Deep Fault zone running roughly down its centre. It offers good evidence that much of the medial eugeoclinal part consists of what may be oceanic material ruckled against the edges of the opposing continents. The Urals Deep Fault zone is itself marked by intense dynamic metamorphism. It includes elongaged blocks of serpentine, gabbro and other basic rocks and is virtually continuous for more than 2000 km. It has every appearance of being the Uralide suture with the basic igneous rocks representing ocean-floor tholeiites. Devonian sediments here are metamorphosed to greenschist facies. The miogeosynclinal zone is about twice as big as its partner and lies to its west (that is, it occurs largely to the west of the Deep Fault zone and includes not only Novaya Zemlya but extends the full length of the western flank of the Urals). It has been very thoroughly examined in the search for oil and gas. The eugeosynclinal zone occurs near the mouth of the River Ob and from around Sverdlovsk to the southern end of the mountains. The general facies arrangements are:

| | (Western flank of Urals) | (Eastern flank of Urals) |
	Miogeosynclinal	Eugeosynclinal
Upper Devonian	Greywackes, sandstones, limestones and slates	Greywackes and slates, minor volcanic, some cherts. Some limestones
Mid-Devonian Lower Devonian	Sandstones, shales and limestones	Sandstones and shales with thick spilites, keratophyres and other volcanics

Ordovician–Silurian continental shelf rocks below the Devonian.

A number of transverse uplifts separate different structure/facies zones within the miogeoclinal western slope of the Urals. Facies changes along strike are locally sharp and conspicuous, reflecting basement control during Devonian times. Locally the carbonates are very thick. In contrast the eastern slope shows a series of volcanic belts with very thick accumulations of Devonian deep-water effusives and sediments alternating laterally with sharp uplifts on which are shallow-water and even continental deposits. Reefs occurred on some uplifted parts of the eugeoclinal floor. Basaltic and andesitic masses several thousand metres thick are especially prominent in the southern reaches of the eastern part of the Urals. Between 2000 m and 3000 m of Devonian rocks are present, being very varied in facies and rich in fauna. Indeed it is said that a standard succession that is comparable to that in Western Europe or North America could be proposed here. Murchison's own travels through parts of the Ural Mountains helped convince him of the validity of the new system between the Silurian and the Carboniferous rocks.

Two limestone facies occur in the Ural Mountains as in the Balkan area. The reef limestones and associated beds of the east Urals can be regarded as Hercynian facies, pale in colour and distinct in fauna from the Rhenish facies to the west. However, the dark and bituminous thin-bedded limestones and shales with a typical pelagic fauna of Givetian to Famennian age reach 200 m or more; they are distinctive and known as the Domanik facies. These are thought by some Soviet geologists to have accumulated in stagnant sea-floor depressions under flotant algal mats or at least where an algal flora was extremely dense. Other workers in Tartaria regard the Domanik facies as typical of submarine uplifts such as the Tartarian arch (Smirnov, 1973). Bauxites at several levels there, capping thin limestone formations, may have developed during phases of active uplift.

It is clear that Siluro–Devonian diastrophism occurred in the Ural belt adjacent to Kazakhstan and the western Sayan area and that in the extreme south deep-water sediments and basic igneous rocks may extend into the Caucasus.

The Central Asiatic Complex

This vast area (Zonenshain, 1973) stretches from the southern end of the Ural Mountains eastwards between the Siberian Platform and the Tarim and North Chinese Platforms. A further belt of geosynclinal rocks borders the southern margin of the latter two platforms, and extends as an early forerunner of the Tethyan Geosyncline, running eastwards from the Caucasus towards the Pamir Ranges and on into China. This oceanic realm, Prototethys, the Rheic or Hercynian Ocean, is physically rugged and difficult of access. Early Palaeozoic geosynclinal basins here were succeeded by Middle Palaeozoic eugeosynclinal belts of different ages, sizes and rates of evolution. There is a Devonian component in virtually all of them and there was also vigorous Middle to Late Devonian volcanic and orogenic activity. Ophiolitic suites are followed by thick flysch sequences and by calc-alkaline volcanics. Despite orogeny, it is clear that ocean-floor material survives and is not greatly transported. Early ocean basins tended to receive 5000 m or more of flysch as in eastern Kazakhstan with a similar thickness of basaltic and andesitic volcanics, suggesting the rapid development of island arcs in Middle Devonian time. Zonenshain (1973) believes that microcontinents may have been surrounded by such areas and were rising vigorously at those times. Devonian volcanics associated with this are especially conspicuous in the southern Gobi area.

At its western end the belt is revealed in deposits in the South of France, the Pyrenees, Spain and Portugal. Some 400 m or more of flysch deposits and black shales occur in the axial zone of the Pyrenees, and are said to have been deposited in a series of narrow, subparallel furrows or troughs, separated by ridges.

Devonian rocks are widespread within the Caucasus and Armenia and the Nakhichevan A.S.S.R. They include members of all three Devonian series, consisting of carbonates and clastics (several thousand metres) and volcanics somewhat metamorphosed and deformed. Eight Palaeozoic geosynclinal and geoanticlinal belts have been claimed in the Caucasus but the geoanticlines are devoid of the Devonian. During Middle Devonian time volcanism was intense. Beyond the U.S.S.R. the belt continues in the broad geosynclinal domain north of the Inshan–Tianshan mountains of China; that is, the Junggar–Hingan region, and in the Kunlun Shan and the South Tianshan region, and the West China–North Tibet–Himalaya belt (Yang *et al.*, 1981). The Hingan area exhibits eugeosynclinal and miogeosynclinal rocks in profusion, while the central and southern areas show the (Baoxin-type) carbonates, clastics and volcanics deposited in a deeper unstable environment with many lateral facies changes. Devonian strata are widespread and largely marine, being thickest in the central ranges. All stages are present and there are thick sequences of volcanic rocks. In general the oldest formations are found in the southern ranges. Thick carbonates include reefs and open-water Hercynian-type facies.

The West Pacific Geosyncline

Although least known of the geosynclinal belts of the U.S.S.R., this nevertheless is one of the most complex and interesting, and is now the subject of much research. To the north-east the belt appears to connect with Alaska and to the south it runs into Japan; throughout much of its length its easternmost and southern-most outcrops reveal 'abyssal' magmatism and volcanism. It has been evolving throughout much of the Phanerozoic era and occupies the region between the Amur and Lena rivers and the Pacific and Arctic coasts. Vladivostock lies at the southern end in the U.S.S.R. Devonian formations here are highly deformed and metamorphosed. Nevertheless fossils are found and have led to the concept of two palaeontological provinces within the geosyncline—the northern being akin to Europe and the southern more obviously having affinities to Kazakhstan, the Kyzbas and China. This 'Asiatic' province or subprovince is distinguished by many endemic species of corals and brachiopods. Nevertheless it was clearly comprised of several discrete marine basins that subsided between volcanic ridges or arcs throughout the Devonian period.

The Samfrau Geosyncline

One of the major geological features of the Late Palaeozoic world was the great arc and trench belt, referred to by Alexis du Toit as the 'Samfrau Geosyncline'. It is represented in Devonian geology by thick rock sequences in South America, South Africa, Antarctica and Australia. Miogeosynclinal accumulation appears to be the mode in the Andean portion of the Samfrau Geosyncline, but all other areas include eugeosynclinal suites as well.

Orogenic movements occurred within this great mobile belt throughout the long existence—the Pampean movements in South America, the Tabberabberan and others in Australia, and, doubtless, equally strong and effective movements took place in the antarctic portion of the geosyncline. Downwarping and neritic deposition appears to have begun in Early Palaeozoic time in virtually all parts of the belt but only in eastern Australia can much of the pre-Devonian evolution of the crust be deciphered. In Late Palaeozoic time the geosyncline was locally affected by subduction-induced orogeny, and all but the Andean area became stabilised. This last-named, however, has undergone plutonism, volcanism and orogeny since then in response to the continuing subduction of the Nazca plate. Mesozoic and later continental drift has separated the remaining segments of the Samfrau and they have remained relatively unaffected by movements other than isostatic.

The Andean Basins

The presence of pre-Devonian complexes in the Andean region of South America is well established and a number of strong phases (or events) of deformation occurred over the period of the Proterozoic and Lower Palaeozoic (see figure 9.10). Although the plate-tectonic interpretation of these features calls for vigorous collisions, it is arguable that the Pacific Ocean existed in mid-Palaeozoic time with subduction along or near much of the Pacific margin of South America. Strong Late Silurian orogenic activity is indicated in the Venezuelan Andes, comparing in age with the Atico event of Peru and a similar one in north-west Argentina and Chile.

The Early Palaeozoic also saw a geosynclinal development off the coastal region of what is now Peru, terminating in an orogenic and metamorphic event referred to as the Marcona. This was followed

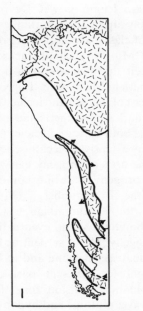
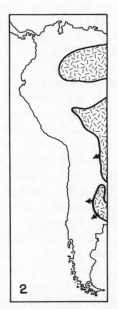
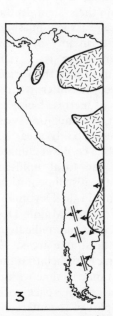
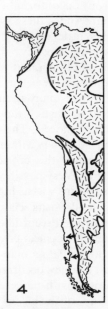

Figure 9.10. The western margin of the South American craton was in Middle Palaeozoic times affected by a series of earth movements in which elongate arcs and uplifts were continuously in process of formation and decay. As they arose they shed great quantities of sediment and volcanic debris into the adjacent actively subsiding and locally deep basins. Sediment provenance is indicated by the arrows in the figures above. 1, Siegenian; 2, Emsian; 3, Late Eifelian; 4, Late Eifelian to Late Givetian (from various sources: Harrington, Copper, and others).

by the intrusion of the large San Nicolas batholith which is dated as around 390 million years old.

Following these events, the Andean region of South America appears to have been the site of three major elongate and migratory troughs during Devonian times. They received a vast load of terrigenous sediment, varying from fine silica shales to coarse conglomerates and olistostrome-like masses of giant blocks. At one time or another they reached from Venezuela in the north to the southern-most parts of Patagonia. All are miogeosynclinal in type, deriving their sediments either from the relatively stable areas to the east or from mobile uplifts within the general Andean region. The major troughs are designated

(1) Venezuelan–Argentina Basin
(2) Precordilleran Basin
(3) Patagonian–Cordilleran Basin

The Devonian was clearly a period of crustal unrest in this part of the world. In several areas thick Ordovician and Silurian terrigenous sedimentary formations underlie the Devonian and there are numerous unconformities within the system. In the Peruvian Andes these systems are largely missing. There is little or no carbonate in any of these successions. Locally, however, there is a rich brachiopod fauna, but for the most part the varied formations in the belt are not very fossiliferous and show in their many thousands of metres thickness and coarse clastic grades that accumulation was very rapid indeed. These were probably areas of very turbid waters with no doubt a cool wet climate that affected the erosion of the local uplifted cordilleras. Very possibly the rate of sedimentation in these basins was as high as any in the Devonian. Glacigene deposits, claimed from the Middle Devonian of the Precordilleran Basin, may indicate a continental ice-sheet cover of the source areas.

It is in the Patagonian section of this belt that the greatest changes occurred from time to time within the period, with fierce erosion keeping pace with subsidence of the basins and uplift of the intervening uplands. Late in the Devonian period these geosynclinal basins were affected by strong earth movements and a low grade of metamorphism was imparted to some rocks. Intrusions of ultrabasic rocks occur in some formations of the Precordillera.

South Africa

In the Cape Belt of South Africa some 3000 m of continental and marine clastic rocks include representatives of the Lower Devonian that strongly resemble the terrigenous and miogeosynclinal Lower Devonian of the Falkland Islands (see figure 9.11). The Bokkeveldt Series (800 m) in the Cape area is dominantly argillaceous and contains a Lower Devonian benthonic fauna. The overlying Witteberg Series (800 m) is more continental in aspect, consisting largely of sandstones and with plants, arthropoda and rare vertebrates.

The Tasman Geosyncline

That great student of geosynclines, Charles Schuchert, first used the name 'Tasman Geosyncline' for the enormous suite of Palaeozoic sediments and igneous rocks in eastern Australia. The eastern highlands and east coast of the continent reveal an extremely thick pile of Devonian rocks infilling parts of the Tasman Geosyncline. In many places throughout its 2500 km length Lower Devonian rocks rest conformably upon the Silurian giving unbroken sequences of clastic and carbonate rocks up to 5000 m thick. 'Miogeosynclinal' facies are found in a western belt that extends from northeastern Queensland south to Melbourne. 'Eugeosynclinal' rocks lie to the east of this belt and include typically greywackes and dark shales, cherts and a wide range of volcanic rocks both basic and acidic. All of these rocks have been extensively tectonised with successive shifts of the area of orogeny eastwards between the Bowning orogeny (Late Silurian–Early Devonian) through the Tabberabberan (Middle Devonian) into the Kanimblan (Late Devonian–Early Carboniferous) and eventually to the Hunter Bowen orogeny. Major serpentine belts occur within the Tasman geosyncline and at least one has been thought to represent part of a Siluro–Devonian (ophiolite) collision zone. Each orogenic belt has effusive, plutonic and hypabyssal intrusive rocks in great abundance and their north–south elongated outcrops serve to emphasise the tectonic grains. Two major geosynclinal regions have been designated within the overall area. The Lachlan Geosyncline extends roughly south from

	SOUTH AMERICA	FALKLAND ISLANDS	SOUTH AFRICA
	(Buenos Aires Prov.) Argentina		
	U. Palaeozoic tillites	U. Palaeozoic tillites	Dwyka tillites
		Port Stanley Series 700 m	Witteberg Series 780 m
		Port Philomel Series 100 m	
			Bokkeveld Series 800 m
	Lolén Fm. 400 m	Fox Bay Series 780 m	
	Providencia Fm. 220 m		Table Mt. Series? (age uncertain)
	Napostá Fm. 370 m	Port Stephens Series 1620 m	
	Bravard Fm. 200 m		
			Precambrian
	Silurian		

(left axis labels: Middle–Upper? Devonian / Lower Devonian; Ventana Group labels in South America and Falkland Islands columns)

Figure 9.11. Southern South America, the Falkland Islands and South Africa each have a thick sequence of predominantly quartzitic sandstone and shale Devonian sediments. Their sedimentological similarities are coupled with palaeontological links, that is, the Lower Devonian Malvinokaffric invertebrate fauna. The Witteberg Series also has Middle Devonian genera of fishes, almost certainly non-marine and unique for this realm, though the Malvinokaffric realm had probably no discrete identity by that time. Although these successions constituted part of the evidence for the 'Samfrau Geosyncline', they cannot be regarded as 'geosynclinal' in anything but the broadest miogeosynclinal sense.

the latitude of Rockhampton, Queensland, to Tasmania; the New England Geosyncline to the east extends from north of Townsville, Queensland, to the coast east of Canberra. The Early Devonian orogeny has stabilised the western part of the Tasman Geosyncline and from the upthrust land mass a flood of molasse was shed to the east. Volcanic activity broke out again, this time from centres east of the old Lachlan site.

A splendid example of Early or Middle Devonian volcaniclastic and flyschoid deposits recently described from New South Wales (Cas *et al.*, 1981)

illustrates the interruption of relatively deep-water mudstone sedimentation by masses of limestone blocks and breccia, sand and volcanic debris (see figure 9.12). Mass flows, debris flows and avalanches are indicated at several levels in a succession that is about 825 m thick.

The Tabbarabberan and Kanimblan orogenies effectively closed the Lachlan Geosyncline and a new cycle of sedimentation subsequently began to the east, continuing into the Carboniferous and Permian. Much of the coarser clastic sediment in the later trough was derived from the recently elevated

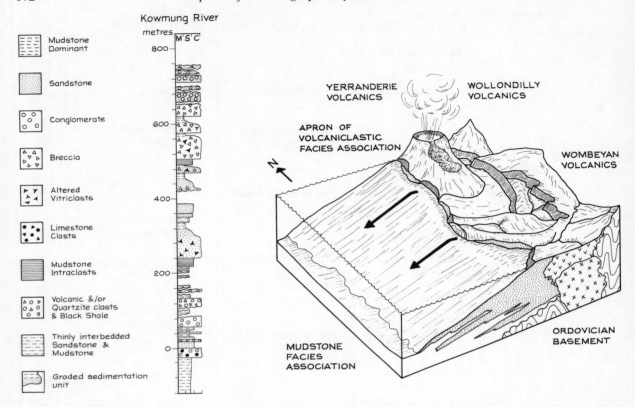

Figure 9.12. The Kowmung Volcaniclastics of New South Wales record the emission and transport of coarse to fine ashes over a submarine volcanic apron and into relatively deep water. Mudstone and sediment, up to boulder-size blocks, were also transported by mass—and turbidity—flows down the same slopes (after Cas *et al.,* 1981).

volcanic uplands on the Lachlan Geosyncline. The New England Geosyncline itself was rich in volcanic rocks and also received open ocean deposits; arc development was conspicuous at the onset of Late Devonian time. Although they outcrop only over a very small area, Devonian rocks in New Zealand appear to correlate with those of Victoria. They were involved in the Late Devonian Tulma orogenic episode in New Zealand.

Throughout its long history the Tasman geosyncline appears to have evolved in a series of eastward zones of accretion in which arcs collided with the continent. The region is so rich in igneous rocks of all kinds that local radiometric data may eventually add much to our understanding of the magmatic and tectonic history.

In summary, deep-water basinal deposition during the Devonian period was widespread and vig-

orous. Oceanic sediments and volcanic arc formations in the 'middle' zones of the time contributed a large part of the infilling of the Middle to Late Palaeozoic geosynclines.

It can be assumed that they represent the deposits of oceanic basins of varying widths which underwent closure and subduction. Late Palaeozoic orogenies brought to an end this process as in eastern America, southern and central Europe and the 'Mediterranean' geosynclinal area, the Urals and the Cordilleran and Andean regions. All manner of other sedimentational evidence indicates that rapid subsidence was accompanied by rapid sedimentation in many eugeosynclinal regions. Thick clastic wedges and prisms in the miogeosyncline indicate vigorous continental uplift and erosion and marginal subsidence. Ophiolitic suites of Middle and Late Palaeozoic age are widespread and there can be little doubt that the 'turnover' of oceanic crust was proceeding at a high rate. Ronov *et al.* (1980) regard the Late Devonian as second only to the Late Triassic as a time of volcanicity, with an average of 10.2 m added to the column per million years, making up about 30 per cent of the total known volume of Phanerozoic rocks. Most of this volcanicity was submarine and geosynclinal. High sea levels and the Late Devonian transgressions may indicate a spread and expansion of the oceanic ridges, and subduction was also clearly almost as vigorous. This is in keeping with the production of the abnormally great volumes of Devonian deep-water basalts that were noted by the Soviet geologists.

Recent deep-sea sediment studies hold great significance for our understanding of sea-level and climatic changes. Comparable studies in Devonian oceanic rocks might reveal significant cycles in sediment deposition related to climate, eustasy etc. As the biostratigraphy of Devonian pelagic organisms in basinal and epicontinental rocks improves, discussion of such phenomena will be that much more plausible. Micropalaeontology in particular may be an effective tool for this.

10

Devonian Geography

In the previous chapters we have examined some of the major associations of sedimentary rocks that are present in the Devonian system and have attempted to relate their origins to models of modern sedimentary environments and processes. This is a rather crude exercise and refinements are highly desirable. Most of the models of Devonian sedimentary environments must stand or fall not only on the basis of their immediate suitability but also on whether or not they are valid in the context of regional, and ultimately, world Devonian palaeogeography. In many cases we are not certain that the models account for every aspect of a particular sedimentary assemblage. For example, the detailed and convincing interpretation of upwards-fining cycles within the Old Red Sandstone facies has been applied to a very large number of formations, which now on closer inspection reveal details that call for a different model. This is not to deny the usefulness of the original concept but it serves to emphasise that interpretations need to be regarded as temporary, to be changed as improvements in knowledge continue. Since most of what is on the previous pages of this book is culled from the literature it is likely to be already out of date, but that is the fate of works such as this.

Can we, nevertheless, address ourselves to some of the questions posed by Hallam (1977), referred to on page 3, insofar as this system is concerned? Advances even since his words were written show that geologists recognise well the aims and necessities he stated. A recent account of Palaeozoic world geographics (Bambach *et al.,* 1981) emphasises that the evolution of the continents of the earth passes from one extreme condition, Pangaea, to another wherein all the continents are dispersed:

the Devonian was intermediate. Here we can offer a view of Devonian geography, perhaps assess the contribution it makes to understanding Phanerozoic earth history, and recognise its limitations and uncertainties.

From the data now available it is possible to build up a picture of the major geographical features of the earth's surface as they were during each of the Phanerozoic epochs. For the geographies of longer spans of geological time the picture has to be more generalised, while for shorter times it may lack local detail in the absence of precise information. Interpretation of the geological (and palaeontological) record improves as our knowledge of the present-day world improves.

The science, or art, of palaeogeography, then, turns upon the uniformitarian axis and palaeoecology is similarly linked to recent ecology and biogeography. Interest in slow or widely separated fluctuations in the evolution of the earth has been rising and the present condition of the planet may be exceptional, even unique, in many ways. However, the modern world serves well enough as a model with which to begin. Its basic patterns of oceanic and atmospheric circulation and interaction of these with land (and mountain) and sea distribution are regular, but far from being a series of latitudinal zones or belts. There remains the problem as to what extent changes on the earth are influenced from elsewhere within the solar system or beyond. Have there been astronomical changes or influences at work which could change our interpretation of the data collected from one geological system or another? Extraterrestrial influence upon the evolution of the planet and of life, and especially upon the sudden mass extinction of taxonomically very different groups, has been a popular field for conjecture and theorising. So far conclusive evidence that any such agency has played a decisive part in the history of life at any particular time is lacking. For the moment we acknowledge that they may have existed but that our first task is to establish uniformitarian models and to ponder extraterrestrial and other influences later.

In attempting a world palaeogeography it may be best to establish the larger features first and progress to smaller regional and local phenomena later. Thus our pattern might be:

1. To establish continental positions and configurations by palaeomagnetic studies.
2. To assess the extent of the epicontinental seas that cover the continental, or cratonic, blocks, and the topographies of the land masses.
3. To establish the nature of the continental margins and of the earth movements that affect them.
4. To postulate oceanic circulation patterns and the climatic belts that are so strongly dependent upon them.
5. To distinguish the major features of the climatic variations within these belts.
6. To assess the major biogeographical features of the day, both in terms of marine invertebrate provinces, vertebrates and continental faunas and floras.

In addition, some consideration may be given to such phenomena as volcanic activity, the length of the day and year, tidal ranges, atmospheric composition and so on. Since the Devonian period lasted perhaps 50 million years, or almost as long as the Cainozoic era, significant changes would have occurred during this time. The nature of these changes and the rate at which they took place are relevant to earth history as a whole.

Continents Located

Devonian palaeogeography can thus be constructed in outline on the basis of such 'firm' lines of evidence as palaeomagnetic data, precise biostratigraphic and isotope chronology, basin and facies analysis, tectonic patterns and igneous activity, palaeoclimatic indicators (evaporites, corals etc.). Local phenomena such as tidal ranges and prevailing wind directions can be determined by a comparison with models of Recent phenomena, and climatic and oceanographic features can be suggested. Several such models of the Devonian world have recently been presented (Morel and Irving, 1978; Heckel and Witzke, 1979; Scotese *et al.*, 1979; Ziegler, A. M. *et al.*, 1979; Smith *et al.*, 1981). Morel and Irving's model is shown in figure 10.1.

It is, of course, necessary to give some consideration to what constitutes a continent, and to what is the nature of a continental margin. Continents today have rather sharp boundaries or outlines, many possessing extensional margins with scarp edges, but outlines may be blurred by deltas and volcanic activity. Where collisions or compressional effects have taken place volcanic activity and accretionary sediment wedges may occur. Collisions in the past between continents of irregular shape might be expected to leave local gaps or hollows between them, and a few have been suggested on the basis of geophysics. So much continental manoeuvring has, however, taken place since mid-Palaeozoic times that the outlines of the continents of those days have been much obscured. A. M. Ziegler *et al.*, (1977), give an extended discussion of how the Mesozoic and Cainozoic regimes have masked mid-Palaeozoic continental architecture.

Palaeomagnetic data, although abundant and quantitative, are not always to be trusted because of subsequent overprinting and other causes (see Morel and Irving, 1978). This is particularly the case with the smaller cratonic regions or 'microcontinents'. At present we also lack sufficient information about most of the Asiatic 'microcontinents' to do more than suggest that they are identifiable by their stratigraphical characteristics and faunal provincialism. Only at the end of the Palaeozoic do the Asian 'microcontinents' all appear to have become fused to the main Laurasian mass. Some of those that have recently been postulated as underlying southern Europe may have during Devonian time become more closely associated with one another, if not actually fused to the southern flank of the European cratonic region. As is often stated, palaeomagnetic studies indicate palaeolatitudes; longitudinal positions remain uncertain, and it is this that in part allows so many versions of palaeogeography for

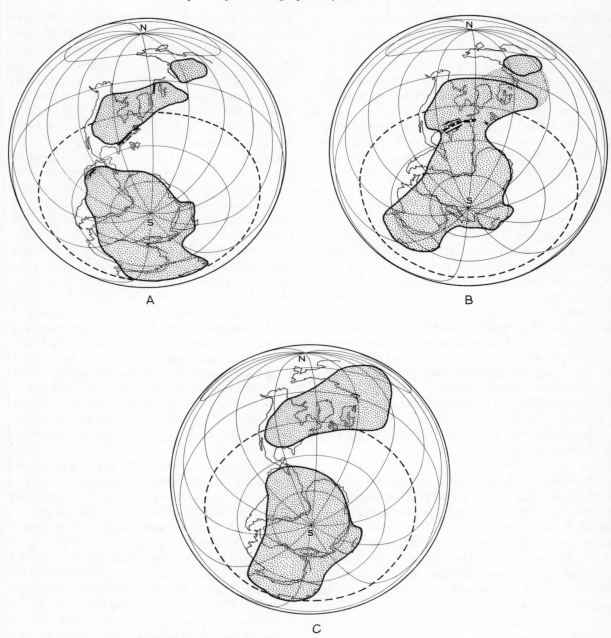

Figure 10.1. Morel and Irving (1978) offered different if less-detailed palaeogeography of Late Silurian to Late Devonian times in three maps, A, B, C (approximately 400–374 million years ago, 375 million years ago, and 370 million years ago respectively) which show the position of the Caledonian and Acadian orogenic belts in collision areas where the northern and southern land masses approach one another. Note that they indicate the British Isles to have been situated far from the Baltic area in pre-Middle Devonian times (A). Their map of Middle Devonian palaeogeography (B) indicates that 'Gondwanaland' rotated sinistrally with respect to the 'Laurasian' mass, so bringing the British Isles into their present position relative to the Baltic Shield. At this point a short-lived 'Pangaea' was thus present, only to split again in Late Devonian times (map C). Note the great range in latitudes—east Asia is close to the polar circle while much of Gondwana lies equally far south (or even farther).

any one time. Morel and Irving (1978) stress that relative longitude is best estimated by minimising relative motions between the continents.

The choice of a suitable map projection is important in palaeogeographic world maps. In attempting to present as conventional a map as possible the Mollweide Projection has the advantage of being an equal area projection as well as having parallel lines of latitude. The Mercator Projection, used in many previous palaeogeographical maps, with its parallel lines of latitude and longitude, results in great distortion of the polar areas. A. M. Ziegler *et al.* (1977, 1979) and Scotese *et al.* (1979) used the Mollweide Projection; Heckel and Witzke (1979) used the similar sinusoidal projection.

The flow of information upon which to base maps that show the positions of the continents in the past is now very great and results in new improved world reconstructions every few years. One of the latest such compilations (Scotese *et al.*, 1979) offers seven reconstructions of the Palaeozoic, Late Cambrian/Middle Ordovician, Middle Silurian, Early Devonian, Early Carboniferous, Late Carboniferous and Late Permian (figure 10.2). The differences between the new reconstructions and those given by other authors recently are not great and although palaeomagnetic determinations are the basic data used in the positioning of the continents' lithofaces, tectonic and other information is also utilised. Both the Mollweide and polar Projections are used to show the globe from both 'front' and 'back' and from north-polar and south-polar viewpoints (see figure 10.3). For these maps the shape and extent of each Palaeozoic continent has to be determined and its orientation with respect to the other continents has to be established. While the shapes of the continents, except Eurasia, have changed relatively little since the break up of Pangaea (200 million years ago), the Palaeozoic continents have been greatly modified by collision or fragmentation. Major suture zones that bound ancient continents are marked by ophiolites, accretionary flysch sequences and other lithofacies, indicating deep seas. For detailed discussion of the evidence used in determining these zones see Ziegler, A. M. *et al.* (1979), Scotese *et al.* (1979), and Burke *et al.* (1977).

Evidence of biogeographical provinces gives a useful check on the predicted positions of continents, though it has to be applied with extreme care (Ziegler, A. M. *et al.*, 1979). Climate, geographic distance and oceanic depth provide the major boundary constraints upon biogeographical provinces and have probably always been effective barriers. Several symposia have considered Palaeozoic biogeography (Middlemiss *et al.*, 1971; Hallam, 1973; Hughes, 1973; Ross, 1974) and only three or four Palaeozoic biogeographic provinces or realms, largely of a latitudinal kind, have normally been suggested.

Gondwanaland

Consisting of South America, Africa, Arabia, Madagascar, India, Australia and Antarctica, Florida, southern and central Europe, Turkey, Afghanistan, Iran and Tibet, and New Zealand, this is the largest and best-known palaeocontinent. Over half the known continental crust is embraced by Gondwanaland, and it has remained remarkably intact since it was stabilised by the Pan-African Event at the end of the Precambrian. Several slightly different assemblages of these elements have been proposed, involving the positions of Madagascar, India and Australia. The one shown in figure 10.3 is perhaps the most generally used. Some authors (Scotese *et al.*, 1979, for example) place Spain, France and Italy south of the mid-oceanic ridge of the Foibic–Rheic Ocean, close to north-west Africa and far distant from south-west England, Belgium and north Germany.

Laurentia, Baltica (North Atlantis)

Laurentia includes the North American craton with Scotland and Ireland north of the Caledonian suture, Greenland and Spitsbergen and the Chukotsk peninsula of north-eastern Siberia. Scotese *et al.* (1979) also treat Mexico, Yucatan and Honduras as separate parts of this assemblage. Baltica consists of the Fenno-Scandian part of northern Europe north of the Hercynian suture and west of the Uralian suture. With the completion of the later Caledonian (Middle–Upper Devonian) earth movements the southern part of the British Isles has become attached to Laurentia and Baltica to form

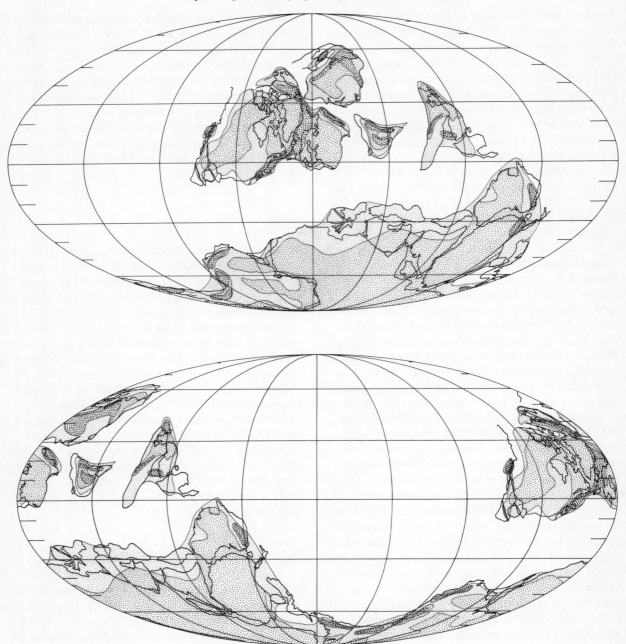

Figure 10.2. An Early Devonian (Emsian) palaeogeography by Scotese *et al.* (1979) that incorporates data from very many sources. The Mollweide Projection provides an excellent framework for the 'northern' continents but produces a severely distorted 'Gondwanaland'. The deep oceans are unshaded, shallow epicontinental seas are shown by light shading, low land by intermediate shading and mountains by heavy stipple.

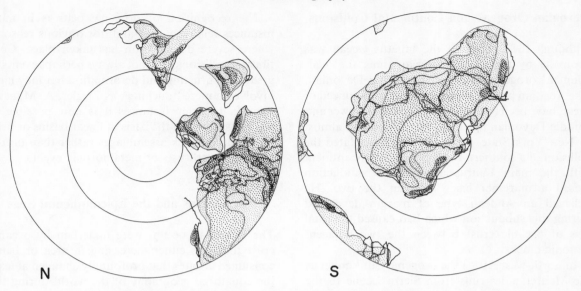

Figure 10.3. Stereographic polar projections of the northern and southern hemispheres, also by Scotese *et al.* (1979), emphasise the presence of 'land hemisphere' and 'ocean hemisphere' in the Early Devonian (Emsian) world. Shading as for figure 10.2.

the supercontinent Laurussia (Euramerica, North Atlantis).

Siberia, Kazakhstania and China

Three major cratonic blocks or palaeocontinents lie to the north of the Himalayas. Siberia was throughout most of the Palaeozoic oriented 180° from its present position. Kazakhstania, lying to the west of the south-west of Siberia, grew throughout the Palaeozoic by the accretion of volcanic arcs and oceanic sediments, finally colliding with Siberia and Baltica in the Late Palaeozoic and producing the eastern part of Laurussia (Scotese *et al.*, 1979).

China really includes several distinct cratonic blocks which ultimately were sutured together in latest Palaeozoic and earliest Mesozoic times. Scotese *et al.* (1979) treat them all as one unit because it is clear that they all belong to the same floral and faunal provinces throughout Palaeozoic time and could not have been widely separated from one another. They also postulate that during Silurian time Baltica moved into the same low latitudes as Laurentia, and Siberia was moving northwards. Several possibilities exist for the movement of

Gondwanaland during this period but it maintained its general antarctic situation throughout.

By mid-Devonian time Laurentia and Baltica had collided and the Caledonian–Franklinian Geosyncline had been eliminated. Siberia (Angaraland) remained as the most northerly of all the cratonic continental elements, while Kazakhstania and the east Asian blocks lay to the east of the Laurentia–Baltica (North Atlantis) mass and relatively separated from it until later Palaeozoic times. Nevertheless, it should be noted that quite different views have been offered by those who find the kaleidoscopic version of Palaeozoic continental distribution hard to accept. Boucot and Gray (1979) argue for a 'Palaeozoic Pangaea'—a single supercontinental mass—and they are able to present much biogeographical evidence in support, not least from the Devonian. To what extent this in turn lends itself to the case for an expanding earth is yet to be seen. Increasing knowledge of the Asian orogenic belts that were active in Devonian time should enable us to assess the extent of some of the crustal shortening and possible kaleidoscopic changes that took place in that part of the 'Laurasian hemisphere' (or of the lack of them).

Devonian Orogeny and Continental Collisions

Although the closure of the Iapetus Ocean was complete by the end of Silurian time, the 'Caledonian' orogenies continued into the Devonian. The Acadian phases, being the most pronounced of these, have been found affecting rocks of Lower and Middle Devonian age from Newfoundland almost to New York State. Keppie (1977) has related the collision of a southern Nova Scotia microcontinent with the main Laurussian mass as subduction closed a marginal basis between the two. He believed an Andean type of margin developed during this subduction which had caused the total loss of oceanic crust between the two adjacent cratonic blocks.

In north-west Africa the orogenic belt known as the Mauretanides runs from Sierra Leone to the Anti-Atlas Mountains. Its origin may be in a Devonian collision between the African craton and the elongate block or microcontinent that is today split between the south-eastern American seaboard and north-western Africa (Arthaud and Matte, 1977).

Several recent interpretations of the evolution of Brittany (see Lefort, 1979) postulate a northern subduction of oceanic crust on its southern side, so drawing the Iberian microcontinent nearer. A collision may have occurred there in the Late Devonian and Early Carboniferous but its exact timing and nature are debatable.

On the northern and western margins of the Laurentian craton the Ellesmerian and Antler orogenies took place in the Franklinian and Cordilleran belts. During late Early Devonian time movement began in the extreme north to affect the Franklinian trough and by the late Middle or early Late Devonian it was felt throughout the entire area of the present Arctic regions (see Johnson, 1971; Trettin, 1973; Christie, 1979). In the Middle and Late Devonian orogenic movement in the western Cordilleran geosynclinal belt continued to a climax at the end of the period, thus corresponding to the span of the Kaskaskia epeiric phase and somewhat following the peak of the Acadian orogeny. In Cordilleran geology its effects involved shifting centres of deposition, local brief uplift, warping and erosion and, eventually, overthrusting eastwards on to cratonic rocks.

The recognition of orogenic belts is in most instances relatively easy and, for obvious reasons, where severe deformation has taken place. Continental collisions were not always perhaps events of great overthrusting and deformation but may have involved a mere 'touching', as Ziegler, A. M. *et al.* remark. In some instances it is now possible to suggest sutures on the basis of linear basins or belts of distinctive rock assemblages rather than on the basis of overthrusts or metamorphic events.

The Oceans and the Epicontinental Seas

The Devonian oceans were underlain by oceanic crust that was either increasing in area or being consumed at rates that profoundly changed at least the equatorial geography of the world during the period. The principal feature of their distribution is a large 'Pacific' hemispheric ocean and relatively narrow latitudinal or longitudinal oceans between the continental masses north of Gondwana. These smaller oceans were largely occluded by the collisions of the 'Hercynian' orogenies and the positions of individual Devonian depositional basins prior to those earth movements are debatable. A good account of the enigmas and dilemmas posed by the Middle and Late Palaeozoic rocks of Mediterranean Western Europe, a most intensely studied region, is given by Rau and Tongiorgi (1981). The width of ocean separating Gondwana and Baltica has been proposed as $30°–40°$ by Scotese *et al.* (1979); Morel and Irving (1978) thought it even bigger while Van der Voo *et al.* (1980) and Tarling (1980) propose as little as $20°$. Tectonic, sedimentological and palaeontological evidence here is not always unambiguous or compatible. That similar difficulties will be encountered as work in the other 'Hercynian' belts continues is highly probable.

The extent to which the continents were flooded during the Devonian period is partly indicated by the nature and disposition of the remaining Devonian sediments upon them. Subsequent denudation has subtracted from this record, and later formations conceal the Devonian over wide areas. Some idea of the areas involved in North America and Eurasia is given by House (1975a, 1975b), in South America by Copper (1977) and in Australia by Brown *et al.* (1968). In the U.S.S.R. Ronov and

Khain (1954) had already made an attempt to assess the area and volume of Devonian formations, but since then new fields of Devonian strata have been discussed in many parts of the world. Ronov *et al.* (1980) give the following updated information for the Devonian:

Oceanic circulation and climates

Major oceanic surface currents arise in response to low-level atmospheric circulation. The flow of the currents may be interrupted by land barriers with anticyclonic gyres formed at low latitudes and

Sea-covered areas (percentage of total area as represented on the present globe)

	Platforms (shelves)		+ Continents (cratonic areas)		+ Geosynclines (slopes)	Duration (million years)
Upper Devonian	19	→	40	→	73	15
Middle Devonian	20	→	41	→	73	16
Lower Devonian	17	→	30	→	50	24

They regard the data for the platforms as the most reliable, since contraction has operated only minimally there since the strata were deposited. These platforms were not covered so extensively again until Late Cretaceous times. It is not clear what is meant here by geosynclines. The difficulties involved in achieving these estimates are many and obvious and the end result is probably an underestimate—non-oceanic parts of the surface of the earth were flooded to almost half their present area. The actual shoreline positions can be determined in many instances and their advance upon the land is chronicled by the attendant palaeontological record. Thus perhaps about 85 per cent of the earth's surface was covered by the seas. The effects of this upon climate, weathering and sediment production must have been strongly marked and the figures given earlier in this book (pages 7, 21, 23) bear this out.

Within the oceanic parts of the surface volcanism was probably widespread, being associated with oceanic ridge extension and expansion. No trace of Devonian undeformed ocean crust remains and the possible configurations of sea mounts and volcanic islands will never be known. It may be imagined that the oceanic hemisphere was in many respects similar to the Pacific hemisphere today in its physiography but unlike the Pacific in the circulation of its waters, there being no strongly longitudinal continental edges within it.

cyclonic gyres at higher latitudes, and the earth's rotation affects oceanic flow so that the strongest currents develop on the western sides of the oceans.

In mid-Palaeozoic times a north-polar circumglobal current, driven by polar easterlies, flowed westward at high northern latitudes. Eastward flow dominated the middle latitudes with maximum strength around 45°N. Anticyclonic gyres may have operated around the northern subtropical oceans. As there were virtually no distinct north–south land barriers in the northern hemisphere in Early Silurian time, western boundary currents would have been of no more than moderate strength.

In southern latitudes the oceanic circulation had three current systems. Two anticyclonic gyres operated in the subtropical oceans and a cyclonic system at higher latitudes over the Gondwana shelf. There would have been a strong western boundary current in each gyre.

The pattern of oceanic circulation is admirably suggested in Heckel and Witzke's (1979) map which is shown in figure 10.4. This shows the probable Middle Devonian pattern and is in accord with what we know of faunal distribution in the Old World, Appalachian and Malvinokaffric Realms.

Palaeoclimatic Clues and Discussions

Uniformitarianism guides our approach to ancient

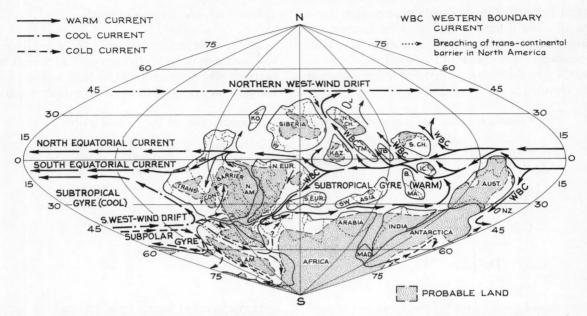

Figure 10.4. By applying the general pattern of present-day oceanic circulation to the configuration of continents in Middle Devonian times, Heckel and Witzke (1979) postulated the model shown here. Klapper and Johnson (1980) found it compatible with the distribution patterns of conodonts and brachiopods. There seems, however, to be only a single conodont realm in offshore facies in the tropics rather than the two realms favoured by the brachiopods.

climates and it is assumed that atmospheric circulation operated in the past on the same principles as at present. The positions of the continents and mountain ranges are thought to have exerted strong influences on the climatic patterns of the past and conspicuous amongst these influences in mid-Palaeozoic time would have been the large Gondwana continent with its centre not far from the South Pole. The circulation pattern there was in consquence more cellular and complex than in the northern hemisphere. A large north-polar ocean would presumably have had a marked ameliorating effect upon the climate within that hemisphere.

A. M. Ziegler *et al.* (1979) raise the important question of the adequcy of the present earth as a climatic model for the Palaeozoic when the distribution of continents and oceans was so different. They note that wind directions are ultimately controlled by processes that are related to differential heating of the earth's surface and to the earth's rotation. Although earth's rotation has slowed with time (Rosenberg and Runcorn, 1975), the Coriolis

forces being stronger and resulting in latitudinal compression of circulation cells at earlier times, only a small change of this kind has occurred since the beginning of the Palaeozoic era (see page 186). Precipitation is largely controlled by the wind belts and the configuration of continents so that idealised models of precipitation patterns for continents with mountain ranges that straddle the wind belts on either the eastern or the western sides of continents may be postulated as shown in figure 10.5 (Ziegler, A. M. *et al.*, 1979, after Robinson, 1973). In these it can be seen that the boundaries of precipitation belts tend to increase in latitude from west to east, a tendency that is accentuated by the mountain ranges.

The equatorial, tropical, temperate and polar climatic belts of the world may be thought of as typified in the production of certain marine sedimentary assemblages. The single equatorial zone is hot and wet, conditions which lead to the production of coals, thick clastic suites and reefs where associated with an active plate margin. Carbonates

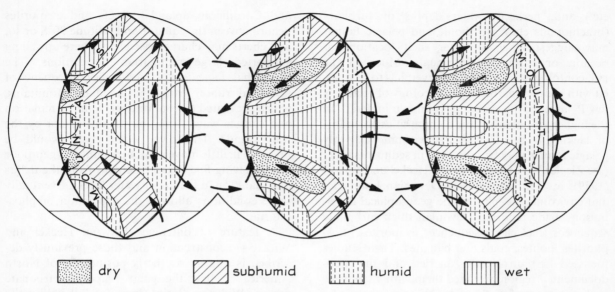

Figure 10.5. An idealised model of world climatic conditions as a function of latitude and Recent geographical configuration offers a basis for postulating those of the past. The model is readily applied to the Devonian continents and the disposition of Devonian climatic indicators seems to be in broad agreement with it. Discrepancies arise possibly because of the narrowness of the seas between the continents and the location of mountain chains (see figure 10.6).

are characteristic of offshore equatorial platformal areas. In the tropical belts (10°–30°) which are warm and dry, evaporites, carbonates and reefs and authigenic minerals are characteristic. Temperate belts (30°–60°) with their cool, variable and rainy climates produce mainly clastic assemblages while the polar zones with cold and dry climates characteristically yield glacial and glacio–marine deposits.

So in many palaeogeographical and palaeoenvironmental essays several litholigical associations, lithofacies and fossils have been specially scrutinised. In addition to those mentioned above, caliches, bauxites and laterites, bedded cherts, aeolian sands, coal or peat are of significance. None is without recognisable limitations as an environmental 'marker'; many accumulate very rapidly and intermittently. The phenomena they represent may be distinctive but ephemeral or impersistent. Plotting the regional or global distributions of these materials promotes but does not necessarily prove conclusively that specific climatic conditions or regimes obtained during the distant past.

A. M. Ziegler *et al.* (1977), for example, studied this kind of evidence for Silurian continental distri-

butions, palaeogeography, climatology and biogeography, and Heckel and Witzke (1979) used similar data for the Devonian. Reliably dated lithofacies provide some of the better base lines, while study of the fossil communities of rocks gives additional evidence of the environment of deposition. A. M. Ziegler *et al.* found it convenient to plot on their maps nine initial lithofacies on the basis of evidence from several hundred individual areas around the world. The major sedimentary types were recorded as clastics, carbonates and mixed clastics and carbonates; minor sedimentary lithologies were evaporites, bedded chert, reefs and authigenic minerals such as chamosite–haematite ores and phosphate nodule beds. They also noted the distribution of different volcanic rocks—andesitic or ophiolitic—as indicative of different tectonic settings, and placed great store on these in their interpretation of continental margins, their movements and evolution. Andesitic suites may be used to indicate the presence of mountains, for example.

Evidence from fossils will give some indication of probable water temperatures (hermatypic corals etc.) and, perhaps, water depths or turbidity (brachiopods etc.). Benthonic and pelagic faunas may suggest inner or outer shelf locations or oceanic or basinal environments. Much of the palaeontological evidence, however, has to be treated with great care, for the autoecology of relatively few Palaeozoic invertebrates is known in detail. In the higher latitudes benthic communities are scarce or lacking from outer shelf environments. Only pelagic faunas may be preserved in sediments there (see Ziegler, A. M. *et al.*, 1979). Heckel and Witzke (1979), in their synthesis of Devonian lithofacies and environments, plotted the geographical distribution of the reported Devonian thick carbonate sequences, thick clastic sequences, evaporites, phosphorites, oolites, coals and bauxites. Their studies focussed first on North America and this well-documented region provided them with a keystone in their picture of world-wide Devonian geography (see figure 10.6). The main features of Devonian lithofacies distribution appear to reflect the subdued and changing relief of the shallowly flooded craton and the presence of mobile belts in the east (Acadian Mountains), the north and west (Franklin and Cordilleran uplands). Reefs and evaporites occur between the equator and latitude 30°S or so, and cherts may characterise areas where upwelling supported large radiolarian populations. The black-shale facies is explained as the consequence of increased runoff from the Acadian Mountains, which prevented normal vertical (water) circulation and surface water being blown out to sea by the trade winds. The lower water layers would be involved in little movement, becoming oxygen-poor and receiving fine organic detritus from the deltas and plankton in the upper water layers. Eventually these conditions allowed the deposition of phosphorite.

A feature of difference between Heckel and Witzke's reconstruction and those previously described is a more southerly positioning of North America so that the great evaporite–carbonate region (Williston Basin) does not fall within the temperate humid zone. A major evaporite sequence—in fact a saline giant—is not likely to develop in an area where there is ample rainfall. The placing of North America in the southern tropics also allows the enormous clastic wedges (and coals)

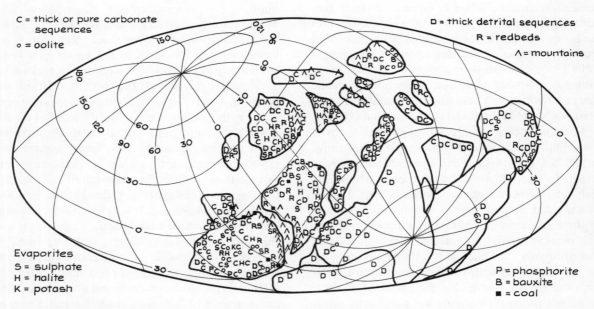

Figure 10.6. Some of the major climatic indicators in the entire Devonian sedimentary array are shown on this suggested Middle Devonian positioning of the continents (after Heckel and Witzke, 1979; and others). They are broadly consistent in conforming to the general climatic regions that are predicted from models of oceanic and atmospheric circulation derived from the Recent (see figure 10.5) and applied to this pattern of continental distribution.

to be derived from the Caledonian–Acadian Mountains in a tropical rain belt.

The Russian Platform similarly has to be located in latitudes where an arid climate prevailed and the very thick clastic and carbonate deposits of both the Urals and central Europe suggest that warm moist climates prevailed. Being on the eastern (windward) side of the tropical Old Red Sandstone Continent, eastern-most Europe felt the effects of trade winds from the Palaeotethys Sea but by the time these reached the western part of the Russian Platform they were dry enough to cause rigorous evaporation. On meeting the Caledonian Mountains these trade winds had to rise and so they lost their moisture again as rainfall. The abundant rivers that drained the mountains eastwards transported great quantities of clastic debris on to the Russian Platform. Where the winds descended on the western side of the mountains they soon lost all their moisture once again and swept westward, being so highly desiccating as to allow potash salts to be deposited in Saskatchewan.

The Siberian continent lying north of the Uralian seaway exhibits also the reefal evaporite–carbonate and red bed successions of a tropical region. Its eastern margin, however, was the site of a north–south chain of uplands or mountains which swung south-westwards into the Salair–Kuznetz area and produced a rain-shadow to the west. In Altai–Sayan, Mongolia and Manchuria thick clastics accumulated on the windward (eastern) side of the volcanic uplands.

Kazakhstan has thick clastics, red beds, coals, bauxites, and carbonates and a great suite of contemporary volcanics, with evaporites and carbonates in the Late Devonian. Between it and the Siberian Angaraland continent lay a narrow geosynclinal seaway. North China and Japan, South China, Tarim and Tibet lay as small cratons to the east and north of the equator. They possess widely varying sedimentary rock successions and volcanic suites, both basaltic and andesitic. It is clear from the faunas and carbonates that their climates ranged from the warm humid to the tropical arid, with volcanic uplands giving local rain-shadows. South China has faunal characteristics that suggest it was separated from the other blocks effectively throughout the Devonian period.

Between Euramerica in the west and Australia in the east the southern tropics contained the microcontinents of southern Europe, south-west Asia, Burma–Malaysia, and Indo-China. They all preserve thick clastic and carbonate sequences and some red beds, with some phosphorites in north and west Iran (south-west Asia). Australia west of the geosynclinal belt along its eastern margin reveals thick detrital sequences, commonly with red beds and local thick carbonates. In south central Queensland a halite evaporitic sequence and in north Western Australia sulphates indicate dry climates.

The North African craton shows a very wide spread of detrital and carbonate sediments. Evaporites are confined to small local occurrences in Morocco and the evidence suggests a position south of 30° for the present North African shoreline, a warm-temperate latitude. India and Pakistan were in a similar position, with thick carbonates being deposited.

Climatic evidence from the more southerly parts of the Gondwanaland supercontinent is not everywhere unequivocal. Detrital successions are locally very thick and monotonous. Evaporites and phosphorites are unknown and carbonates are restricted and of a highly skeletal kind. Cold or wet-temperate climates are indicated with possible glacial activity in parts of eastern South America.

The view of Ziegler, A. M. *et al.* (1977) is eastward travelling low-pressure systems bringing moderate amounts of rainfall to the north coast of Angara, especially during winter, in Silurian times. Dry areas perhaps occurred around 25–30°N, especially on the western sides of the land areas. An intertropical zone where trade winds converged probably brought rain to the eastern slopes of the mountains of Laurentia with a rain-shadow effect on the western slopes. Subtropical high pressure between 25° and 30°S would have produced arid conditions over the Laurentian shelf and the Australian shelf off the west coast of Gondwanaland. Over the Gondwana shelf itself there was probably moderate rainfall throughout the year with summer monsoon rains in the island arc mountains off Australia and Antarctica. Near the pole a dry cold desert would have prevailed but moisture penetrating to the polar latitudes would have been fixed as glacial ice.

In summary, then, it may be noted that, as in the Silurian period, atmospheric circulation during the

Devonian was one in which high-pressure and low-pressure regions were influenced by the positions of the major land masses relative to the normal wind belts. Wind directions were no doubt at low angles across the isobars towards lower pressures. The predominantly oceanic northern hemisphere would probably have experienced a relatively zonally uniform climate. In the southern hemisphere the circulation pattern was, as has been said, more cellular and complex. Diurnal and seasonal changes in the 'weather' rather than in 'climate' may have been much like those of the present. However the length of the Devonian day and year may have been different enough from the present to have led to an overall more equable form of climate.

Devonian Days and Years

About twenty years ago a simple means of finding out something about the Devonian calendar was achieved by J. W. Wells (1963, 1970) of Cornell University. He noted that modern corals produce diurnal increments of carbonate, and counting the carbonate layers gives a close approximation of the age of the coral. The method is reminiscent of finding the age of trees from tree rings. Most corals have about 360 layers (or rings) per year and there may be seasonal differences in the increments. T. Ma (1933, 1934) had suggested that variations in Palaeozoic rugose coral growth might correlate with changes in water temperature (that is, with the seasons) and that those corals growing farthest from the equator experienced the greatest variations. Corals could be used as indicators (relatively) of latitude. Wells, however, examined well-preserved Palaeozoic rugose corals and found that in the Devonian corals 400 growth rings seem to occur in the annual cycle; in Permian forms the figure was 387. His eventual results led to the now well-known graph of growth lines/day set against isotopic dates. Scrutton (1965, 1978) in Britain confirmed Wells' results with other coral material; other workers examined different groups of calcareous organisms, principally molluscs and stromatolites. Scrutton also showed that the Devonian synodic month had 30.6 days (see figure 10.7). A study of the growth lines in the couch chambers of nautiloids tends to confirm the

implications of the coral data (Kahn and Pompea, 1978) and to suggest that the moon was at about half its present distance from the earth. Such a regime would have profoundly affected the tides that occurred in the latitudinal seas between Gondwanaland and the equatorial continents.

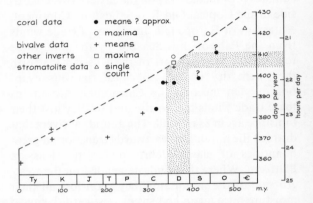

Figure 10.7. The length of the day and number of days in the year throughout Phanerozoic time is thought to be indicated by increments to the calcareous skeletons of various organisms as shown. The dashed line is based on a lengthening of the day by 2.5 ms century^{-1} (after Scrutton, 1978). Discussions of the astronomical implications reveal that in mid-Devonian (375 million years) time the moon may have been at about half its present distance from the earth, and the consequences of that (such as tidal regimes) are not fully understood yet.

The gradual decrease in the number of days in the year and the length of the days has been confirmed, and it is in agreement with the reasoning of geophysicists and astronomers who estimate that there has been a slowing down of the earth's rotational speed over the geological periods (Runcorn, 1964; Broche and Sundermann, 1978). This deceleration is presumably due to tidal friction and amounts to 2 seconds per 100 000 years. The Russian, V. Kotelnikov, calculated in the mid-1960s that between March 1963 and October 1965 the rotation of the earth slowed down so much that the average day lengthened by 1.6 milliseconds. From solar and lunar eclipse records over nearly 2500 years it was suggested that days have been lengthening by an average of 1.8 milliseconds per century. The solar and lunar influence causing tidal drag on the earth's rotation may be modified by the state of the planet itself. It has been thought that small changes in the condition of the earth's core and mantle can influence the rate of slowdown, and that perhaps even such temporary affairs as polar ice caps can influence the rate. Variations in the speed of rotation and in the rate of deceleration seem to be indicated from the geological and palaeontological record. There have even been periods when the rotation seems to have speeded up (Creer, 1975; Kahn and Pompea, 1978). Sudden and repeated reversals in the earth's magnetic polarity ('flips') occur when the speed changes. Whyte (1977) noted the close correspondence between the variation in the rate of rotation of the earth, the polarity bias (normal or reversed) of the geomagnetic field, and the kind of activity at the oceanic ridges, sea level and climate. He accepted Creer's suggestion that there have been periods in earth history when the rate of rotation accelerated and that these have alternated with periods of deceleration. Times when the spin of the earth abruptly changed from accelerating to slowing include the end of the Cretaceous (65 million years ago) and mid-Late Devonian (375 million years ago), and the reverse occurred at the end of the Permian (235 million years ago) and in the Late Ordovician (445 million years ago). The reasons for these changes are far from clear—perhaps they include tidal friction; however, they appear to be complex and need not concern us here. Whyte (1977) and many other workers have maintained that there is a correlation between the turning points, as the times when rotational rate abruptly changed are called, and magnetic reversals and mass extinctions. Magnetic reversals are, however, known to be very much more numerous than the turning points and more significant is the *predominance* of normal or reversed polarity during any period. Correlations between climatic changes and changes in the earth's magnetic field have been sought and demonstrated by several authors so there may well be an equally plausible connection between climate and rotation rate.

The Devonian Period seems to have been predominantly one of frequent geomagnetic polarity reversal, the bias being towards reversed polarity with the Late Devonian especially characterised by that state (McElhinney, 1978). Mass extinctions and periods of high reversal frequency then have

been related in the minds of several writers, but there have been equally well-argued denials of such correlations. Certainly the pattern of such extinctions seems to show the greatest effects amongst the most specialised and diversified faunas, as amongst the offshore benthonic communities and especially the reef communities. Whyte suggests that the climatic instability at a turning point may have been the trigger which, by causing a drop in primary production or a few extinctions, set off a progressive collapse of entire ecosystems. The reduction in climatic gradients during the Silurian and Devonian advocated by Boucot (1975) presented the tabulate coral and stromatoporoid reef assemblages with optimum conditions that were sharply terminated by slight climatic change in the mid-Late Devonian. Linked with the termination of the transgressive character of the seas and a sudden regression, a powerful combination of changes might have been effected to the detriment of many Late Devonian ecosystems.

Mass extinction is evident at the Frasnian–Famennian boundary and was most effective amongst the reefal and peri-reefal organisms. Some 30 per cent of the families hitherto present were lost, mostly those of corals, stromatoporoids, bryozoa, brachiopods, ammonoids, trilobites and fish (see figure 1.12). These are creatures of the relatively shallow waters and many were pelagic. Other pelagic organisms, however, were not affected, the conodonts being one such important group. The reasons for the survival of these groups so far elude us.

This is not to say that we cannot distinguish in the record between groups that are likely to be rapidly affected by environmental change and those that are not. Specialised types with low rate of reproduction and late maturation are commonly the inhabitants of stable environments; those that mature rapidly, and produce many offspring frequently, can move into newly available environments and dominate the early stages of an ecological succession. The former groups may be expected to fare worse than the latter when sharp environmental changes are afoot, and the fossil record confirms this tendency. However, some rapidly evolving groups, the ammonoids being a good example, seem also to have been at risk whenever such changes occurred. Seeking the causes for these phases of mass extinction is difficult but it seems that universal, world-wide causes or factors rather than local events must be involved. There are many good discussions of this topic—Hallam (1981) recently provided one such—but none concludes that a single cause was responsible in each instance within the Phanerozoic. Most probably there was a critical combination of circumstances (spread over a short period of time) involving sea-level changes, orogeny and continental configuration (with *perhaps* some extraterrestrial attendant event). Exactly what the combination was in Late Devonian time remains uncertain. The impact of a giant meteorite as suggested by McLaren (1970) may seem fanciful but it was a well-constructed hypothesis. Alvarez *et al.* (1980) postulated that huge meteorite impacts may have created climatic havoc at the end of the Cretaceous time and brought about mass extinctions then. Direct evidence of any similar Devonian impact event is lacking, but the possibilities of such an event having occurred are real enough.

Having attempted a reconstruction of the major features of the Devonian world, we are left with the need to identify a few of the more important gaps and uncertainties in our knowledge. We can broadly estimate the extent of the Devonian epicontinental seas and we note that many epicontinental sequences seem to show a degree of cyclicity or rhythms. Locally, tectonism may be responsible but it seems inescapable that broad eustatic changes of level are involved. Ocean levels and ocean current systems may both have been influenced by the volcanic activity of the mid-oceanic ridges. From various lines of evidence it is clear that the period was one of impressive orogeny and continental plate movement, so it is probable that mantle convection and active mid-oceanic ridge growth was taking place. The precise dating of sea-level changes, orogenic and tectonic episodes and their resultant sedimentation remains a prime objective if we are to understand how, or even if, a basic plate-tectonic process was at the root of these events. Devonian sedimentation included the widest range of products, and presumably processes, and involved the living world to an unprecedented extent. Environmental controls on organic distributions were no doubt effective but changing during Devonian time. A general synopsis of Devonian palaeobiogeography was given in chapter 5. Tem-

perature gradients alone may have been adequate to have isolated the Malvinokaffric realm and, as Klapper and Johnson (1980) noted, oceanic circulation and the possible presence of land barriers and hypersaline waters may have been as effective as oceanic current systems in the delineation of the Old World and the Appalachian (or Eastern Americas) realms (see figure 10.8). The distribution of vertebrates (Young, 1981) was also more than likely to have been influenced by short-lived islands and connecting arcs between the major continental blocks. The important relative isolation of early Devonian continental 'homelands' for distinctive vertebrate groups soon broke down. By Late Devonian time the vertebrate faunas, still apparently confined to tropical and subtropical regions, became cosmopolitan, following the example of many marine invertebrate faunas.

Meanwhile the evolution of complex terrestrial floras took place with concurrent spread across many continental areas and some geographical variation of the contemporary flora occurred (Chaloner and Sheerin, 1979; Edwards, 1980). The obvious consequences for terrestrial faunas, however, remain tantalisingly poorly recorded (Rolfe, 1980).

The extinction of so many distinctive Devonian vertebrate and invertebrate groups at or near the end of the period remains puzzling. It is tempting to seek an explanation in the culmination of the long transgressive regime of the epicontinental seas and the sudden Frasnian–Famennian regression. Orogenic events at continental margins and the 'beginning of the end' of the Kaskaskia transgression may be linked. And the possibility of the extraterrestrial impact, or whatever, cannot be ruled out as a possible cause of these extinctions.

Each aspect of the study of a geological system is to a greater or lesser extent dependent upon the others and, similarly, advances in the understanding of any or all of these aspects in one system has implications for the other systems. Thus the remarkable increase in our knowledge of the Carboniferous (Mississippian) and Silurian systems in recent years has stimulated and enhanced work on the Devonian. The palaeogeographical model that emerges is still full of uncertainties and creaky hypotheses but its features have been delineated with a uniformitarian

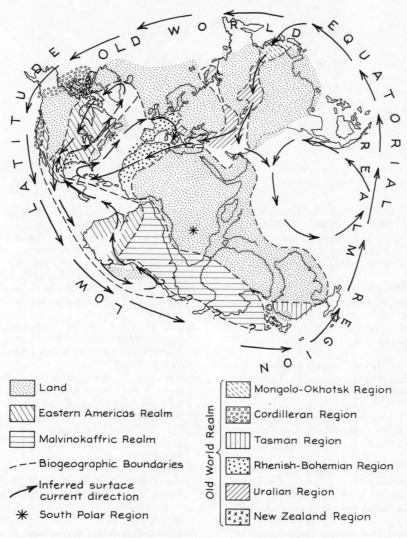

Land

Eastern Americas Realm

Malvinokaffric Realm

- - - Biogeographic Boundaries

Inferred surface current direction

✳ South Polar Region

Old World Realm

Mongolo-Okhotsk Region

Cordilleran Region

Tasman Region

Rhenish-Bohemian Region

Uralian Region

New Zealand Region

Figure 10.8. A Pangaeic reconstruction for the later Early Devonian by Boucot and Gray (1979) postulates a much closer setting of the continents than do those illustrated elsewhere in this chapter and its principal appeal for its authors is the clear relationships of the great faunal realms (and the subdivisions of the Old World Realm) to a simple world geography. Six faunal subdivisions of the Old World Realm are indicated. There is also no need in this model to postulate the migration of neritic benthos across, presumably deep water, oceanic basins to any extent.

bias. During the fifty million years or so of Devonian time there were great geological and biological changes. Many of these can be visualised in terms of models that were not possible twenty-five years ago. The models themselves are likely to be replaced in another decade or two by new and more complex affairs. Most probably the Devonian world will then seem even more different and remarkable.

References

Ager, D. V. (1973). *The Nature of the Stratigraphical Record,* 2nd edition (1981), Macmillan, London.

Alberti, H. (1979). 'Devonian trilobite biostratigraphy'. In M. R. House, C. T. Scrutton and M. G. Bassett (Eds), *The Devonian System. Spec. Pap. Palaeont. No. 23,* 313–324.

Allen, J. R. L. (1964). 'Studies in fluviatile sedimentation: Six cyclothems from the Lower Old Red Sandstone, Anglo-Welsh Basin'. *Sedimentology,* Vol. 3, 163–198.

Allen, J. R. L. (1965). 'A review of the origin and characteristics of recent alluvial sediments'. *Sedimentology,* Vol. 5, 89–91.

Allen, J. R. L. (1974). 'Studies in fluviatile sedimentation: implications of pedogenic carbonate units, Lower Old Red Sandstone, Anglo-Welsh outcrop'. *Geol. J.,* Vol. 9, 1–16.

Allen, J. R. L. (1977). 'Wales and the Welsh Borders'. In M. R. House *et al.* (Eds), *A correlation of the Devonian rocks in the British Isles. Geol. Soc. Lond. Spec. Rep. No. 8,* 40–53.

Allen, J. R. L. (1979). 'Old Red Sandstone facies in external basins, with particular reference to southern Britain'. In M. R. House, C. T. Scrutton and M. G. Bassett (Eds), *The Devonian System. Spec. Pap. Palaeont. No. 23,* 65–80.

Allen, J. R. L. and Friend, P. F. (1968). 'Deposition of the Catskill Facies, Appalachian Region: With Notes on Some Other Old Red Sandstone Basins'. In *Geol. Soc. Am. Spec. Pap. 106,* 21–74.

Allen, J. R. L. and Tarlo, L. B. H. (1963). 'The Downtonian and Dittonian Facies of the Welsh Borderland'. *Geol. Mag.,* Vol. 100, 129–155.

Allen, J. R. L. and Williams, B. P. J. (1981a). 'Sedimentology and stratigraphy of the Townsend Tuff Bed (Lower Old Red Sandstone) in South Wales and the Welsh Borders'. *J. Geol. Soc.,* Vol. 138, 15–29.

Allen, J. R. L. and Williams, B. P. J. (1981b). '*Beaconites antarcticus:* a giant channel-associated trace fossil from the Lower Old Red Sandstone of South Wales and the Welsh Borders'. *Geol. J.,* Vol. 16, 255–269.

Allen, J. R. L., Dineley, D. L. and Friend, P. F. (1968). 'Old Red Sandstone Basins of North America and Northwest Europe'. In D. H. Oswald (Ed.), *International Symposium on the Devonian System, Calgary, 1967, I,* 69–98.

Alvarez, L. W., Alvarez, W., Ansaro, F. and Michel, H. V. (1980). 'Extraterrestrial cause for the Cretaceous–Tertiary extinctions: Experiment and theory'. *Science,* Vol. 208, 1095–1108.

Amos, A. (1954). 'Estructura de las formaciones palaeozoicas de La Rinconada'. *Asoc. Geol. Arg. Rev.,* Vol. 9, 5–38.

Antia, D. D. J. (1980). 'Sedimentology of the type section of the Upper Silurian Ludlow–Downton Series Boundary at Ludlow, Salop, England'. *Mercian Geol.,* Vol. 7, 291–321.

Aronova, S. M., Gassanova, I. G., Krems, A. Y., Lotsman, O. A., Liashenko, I. A., Maximov, S. P., Nechitailo, S. K., Pistrak, R. M., Radionova, A. Ph. and Sokolova, L. I. (1968). 'Devonian of the Russian Platform'. In D. H. Oswald (Ed.), *International Symposium on the Devonian System, Calgary, 1967, I,* 379–396.

Arthaud, F. and Matte, P. (1977). 'Late Palaeozoic strike-slip faulting in southern Europe and northern Africa: Result of a right-lateral shear zone between the Appalachians and the Urals'. *Bull. Geol. Soc. Am.,* Vol. 88, 1305–1320.

Aubouin, J. (1965). *Geosynclines: Developments in Tectonics,* Elsevier, Amsterdam.

Ball, H. W., Dineley, D. L. and White, E. I. (1961). 'The Old Red Sandstone of Brown Clee Hill and the adjacent area. I. Stratigraphy'. *Bull. Br. Mus. Nat. Hist. Geol.,* Vol. 5, 177–310.

Bally, A. W. and Snelsdon, S. (1980). 'Realms of subsidence'. In A. D. Miall (Ed.), *Facts and Principles of World Petroleum Occurrence. Can. Soc. Petrol. Geol. Mem. 6,* 9–94.

Bambach, R. K., Scotese, R. and Ziegler, A. M. (1981). 'Before Pangea: The Geographies of the Paleozoic World'. In B. J. Skinner (Ed.), *Paleontology and Paleoenvironments,* William Kaufmann, Los Altos, California, pp. 116–128.

Bandel, K. (1974). 'Deep-water limestones from the Devonian–Carboniferous of the Carnic Alps, Austria'. In K. J. Hsu and H. C. Jenkyns (Eds), *Pelagic sediments*

190

on land and under the sea. Spec. Pub. Int. Ass. Sedimentol. 1, 93–115.

Banks, H. P. (1980). 'Floral Assemblages in the Siluro–Devonian'. In D. L. Dilcher and T. N. Taylor (Eds), *Biostratigraphy of Fossil Plants*, Dowden, Hutchinson and Ross, Stroudsburg, Pennsylvania, pp. 1–24.

Barrell, J. (1917). 'Rhythms and the measurement of geologic time'. *Geol. Soc. Am. Bull.*, Vol. 28, 745–904.

Beck, C. B. (1960). 'The identity of *Archaeopteris* and *Callixylon*'. *Brittonia*, Vol. 12, 445–456.

Belousov, V. V. (1966). 'Modern concepts of the structure and developments of the earth's crust and the upper mantle of continents'. *Qt. J. Geol. Soc. Lond.*, Vol. 122, 293–314.

Bernard, H. A. and Major, C. F. (1963). 'Recent meander belt deposits of the Brazos River: an alluvial sand model' (Abstract). *Am. Ass. Petrol. Geol. Bull.*, Vol. 47, 350.

Berry, W. B. N. and Wilde, P. (1978). 'Progressive ventilation of the oceans — an explanation of the Lower Palaeozoic black shales'. *Am. J. Sci.*, Vol. 278, 257–275.

Blatt, H. and Jones, R. L. (1975). 'Proportions of Exposed Igneous, Metamorphic and Sedimentary Rocks'. *Geol. Soc. Am. Bull.*, Vol. 86, 1085–1088.

Bott, M. H. P. (1976). 'Mechanics of basin subsidence — an introductory review'. *Tectonophysics*, Vol. 36, 1–4.

Bott, M. H. P. (1980). 'Mechanisms of subsidence at passive continental margins'. In *Dynamics of Plate Interiors. Geodynamics Series 1, Am. Geophys. Union*, pp. 27–35.

Bott, M. H. P. (1982). 'Origin of the lithospheric tension causing basin formation'. *Phil. Trans. R. Soc., Ser. A*, Vol. 305, 314–324.

Bouček, B. (1964). *The Tentaculites of Bohemia: Their Morphology, Taxonomy, Ecology, Phylogeny and Biostratigraphy*, Czechoslovak Academy of Sciences, Prague.

Boucot, A. J. (1960). 'Lower Gedinnian Brachiopods of Belgium'. *Mem. l'Inst. Geol. l'Univ. Louvain*, Vol. 22, 283–324.

Boucot, A. J. (1975). *Evolution and Extinction Rate Controls*, Elsevier, Amsterdam.

Boucot, A. J. and Gray, J. (1979). 'Epilogue: A Paleozoic Pangaea?' In J. Gray and A. J. Boucot (Eds), *Historical Biogeography, Plate Tectonics and the Changing Environment*, Oregon State University Press, Corvallis, Orgeon, pp. 465–484.

Boucot, A. J. and Gray, J. (1980). 'A Cambro–Permian Pangaeic model consistent with lithofacies and biogeographic data'. In D. W. Strangway (Ed.), *Geol. Ass. Canada, Spec. Pub. No. 20*, 389–419.

Boucot, A. J., Johnson, J. G. and Talent, J. A. (1968). 'Lower Middle Devonian faunal provinces based on Brachiopoda'. In D. H. Oswald (Ed.), *International Symposium on the Devonian System, Calgary, 1967, II*, 1239–1254.

Boucot, A. J., Johnson, J. G. and Talent, J. A. (1969). *Early Devonian brachiopod zoogeography. Geol. Soc. Am. Spec. Pap. 119*.

Boucot, A. J., Dewey, J. F., Dineley, D. L., Fletcher, R., Fyson, W. K., Griffin, J. G., Hickox, C. F., McKerrow, W. S. and Ziegler, A. M. (1974). *Geology of the Arisaig Area, Antigonish County, Nova Scotia. Geol. Soc. Am. Spec. Pap. 139*.

Boucot, A. J., Gray, J., Fang, R-S., Yang, X-C., Li, Z-P. and Zhang, N. (1982). 'Devonian calcrete from China: its significance as the first Devonian calcrete from Asia'. *Can. J. Earth Sci.*, Vol. 19, 1532–1534.

Bouma, A. H. (1962). *Sedimentology of Some Flysch Deposits: A Graphic Approach to Facies Interpretation*, Elsevier, Amsterdam.

Bowen, Z. P., Rhoads, D. C. and McAlester, A. L. (1974). 'Marine benthic communities in the Upper Devonian of New York'. *Lethaia*, Vol. 7, 93–120.

Breivel, M. G., Eroshevskaya, R. I., Nestoyanova, O. A. and Khodalevich, A. N. (1968). 'Devonian of the eastern slope of the Urals'. In D. H. Oswald (Ed.), *International Symposium on the Devonian System, Calgary, 1967, I*, 421–432.

Briggs, L. I. (1958). 'Evaporite facies'. *J. Sedim. Petrol.*, Vol. 28, 46–56.

Broche, P. and Sundermann, J. (1978). *Tidal Friction and the Earth's Rotation*, Springer-Verlag, New York.

Brown, D. A., Campbell, K. S. W. and Crook, K. A. W. (1968). *The Geological Evolution of Australia and New Zealand*, Pergamon, Oxford.

Brynhi, I. (1978). 'Flood deposits in the Hornelen Basin, west Norway (Old Red Sandstone)'. *Norsk Geol. Tidssk.*, Vol. 58, 273–300.

Brynhi, I. and Skjerlie, F. J. (1975). 'Syndepositional tectonism in the Kvamshesten district (Old Red Sandstone), western Norway'. *Geol. Mag.*, Vol. 112, 593–600.

Burchette, T. P. (1981). 'European Devonian Reefs: A Review of Current Concepts and Models'. In D. F. Toomey (Ed.), *European fossil reef models. Spec. Pub. Soc. Econ. Paleont. Miner., Tulsa*, No. 16 85–142.

Burke, K., Dewey, J. F. and Kidd, W. S. F. (1977). 'World distribution of sutures — sites of previous oceans'. *Tectonophysics*, Vol. 40, 69–99.

Burrett, C. F. (1974). 'Plate Tectonics and the Fusion of Asia'. *Earth Planet. Sci. Letters*, Vol. 21, 181–189.

Cant, D. J. and Walker, R. G. (1976). 'Development of a braided-fluvial facies model for the Devonian Battery Point Sandstone, Quebec'. *Can. J. Earth Sci.*, Vol. 13, 102–119.

Carey, S. W. (1975). 'The Expanding Earth — an Essay Review'. *Earth Sci. Rev.*, Vol. 11, 105–143.

Cas, R. (1979). 'Mass flow arenites from a Palaeozoic interarc basin, New South Wales, Australia'. *J. Sedim. Petrol.*, Vol. 49, 29–44.

Cas, R. A., Powell, C. McA., Fergusson, C. L., Jones, J. G., Roots, W. D. and Fergusson, J. (1981). 'The Lower Devonian Kowmung Volcaniclastics: a deep-water succession of mass-flow origin, north eastern Lachlan Fold Belt, N.S.W.', *J. Geol. Soc. Australia,* Vol. 28, 271–288.

Chaloner, W. G. and Richardson, J. B. (1977). 'Southeast England'. In M. R. House (Ed.), *A correlation of Devonian rocks of the British Isles. Geol. Soc. Lond. Spec. Rep. No. 7,* 26–39.

Chaloner, W. G. and Sheerin, A. (1979). 'Devonian Macrofloras'. In M. R. House, C. T. Scrutton and M. G. Bassett (Eds), *The Devonian System. Spec. Pap. Palaeont. No. 23,* 145–161.

Chibrikova, E. V. and Naumova, S. N. (1974). 'Zonal complexes of Devonian spores and pollen in the European part of U.S.S.R.'. *Proc. 3rd Int. Palynol. Conf. Novosibirsk, 1971,* Nauka, Moscow, pp. 39–47 (in Russian).

Chlupáč, I. (1968). 'The Devonian of Czechoslovakia'. In D. H. Oswald (Ed.), *International Symposium on the Devonian System, Calgary, 1967, I,* 109–126.

Chlupáč, I. (1976). 'The Bohemian Lower Devonian stages and remarks on the Lower–Middle Devonian Boundary'. *Newsl. Stratig.,* Vol. 5, 168–189.

Chlupáč, I., Jaeger, H. and Zikmundova, J. (1972). 'The Silurian–Devonian Boundary in the Barrandian'. *Can. Petrol. Geol. Bull.,* Vol. 20, 104–174.

Chlupáč, I., Lukeš, P. and Zikmundova, J. (1977). *Field Conference of the Int. Subcomm. on Devonian Stratigr. Barrandian 1977. A Field Trip Guidebook,* Geological Survey, Prague.

Christie, R. L. (1979). 'The Franklinian Geosyncline in the Canadian Arctic and its relationship to Svalbard'. *Saert. Norsk Polarinst. Skrifter,* Vol. 167, 263–314.

Churkin, M. (1974). 'Paleozoic marginal ocean basins—volcanic arc systems in the Cordilleran fold belt'. In *Modern and Ancient Geosynclinal Sedimentation, S.E.P.M. Spec. Pub. 19,* 174–192.

Clark, T. H. and Stearn, C. W. (1960). *The Geological Evolution of North America,* The Ronald Press Co., New York.

Clayton, G., Graham, J. R., Higgs, K., Holland, C. H. and Naylor, D. (1980). 'Devonian rocks in Ireland: a Review'. *J. Earth Sci. R. Dubl. Soc.,* Vol. 2, 161–183.

Conybeare, C. E. B. (1979). *Lithostratigraphic Analysis of Sedimentary Basins,* Academic Press, London.

Cooper, G. A., Butts, C., Caster, K. E., Chadwick, G. H., Goldring, W., Kindle, E. M., Kirk, E., Merriam, C. W., Swartz, F. M., Warren, R. S., Warthin, A. S. and Willard, B. (1942). 'A correlation of the Devonian sedimentary formations of North America'. *Bull. Geol. Soc. Am.,* Vol. 53, 1729–1794.

Copper, P. (1966). 'Ecological distribution of atrypid brachiopods'. *Palaeogeogr. Palaeoclimatol. Palaeoecol.,* Vol. 2, 245–266.

Copper, P. (1977). 'Palaeolatitudes in the Devonian of Brazil and the Frasnian–Fammenian mass extinction'. *Palaeogeogr. Palaeoclimatol. Palaeoecol.,* Vol. 21, 165–207.

Creer, K. M. (1975). 'On a tentative correlation between

changes in geomagnetic polarity bias and reversal frequency and the earth's rotation through Phanerozoic time'. In G. D. Rosenberg and S. K. Runcorn (Eds), *Growth Rhythms and the History of the Earth's Rotation,* Wiley, London, pp. 293–318.

Crimes, T. P. and Harper, J. C. (Eds) (1970). *Trace Fossils,* Liverpool Geol. Soc., Seel House Press, Liverpool.

Crosby, E. J. (1972). 'Classification of Sedimentary Environments'. In J. K. Rigby and W. K. Hamblin (Eds), *Recognition of ancient sedimentary environments. Spec. Pub. Soc. Econ. Paleont. Miner., Tulsa, No. 16,* 4–11.

Cuffey, R. J. and McKinney, F. K. (1979). 'Devonian Bryozoa'. In M. R. House, C. T. Scrutton and M. G. Bassett (Eds), *The Devonian System. Spec. Pap. Palaeont. No. 23,* 307–312.

Dawson, J. (1862). 'On the flora of the Devonian period in north-eastern America'. *Q. J. Geol. Soc. Lond.,* Vol. 18, 296–330.

Dewey, J. M. and Bird, J. M. (1970). 'Plate tectonics and geosynclines'. *Tectonophysics,* Vol. 10, 625–638.

Dickinson, W. C. (Ed.) (1974). *Tectonics and sedimentation. Spec. Pub. Soc. Econ. Paleont. Miner., Tulsa, No. 22,* 1–24.

Dietz, R. S. and Holden, J. C. (1966). 'Miogeoclines in space and time'. *J. Geol.,* Vol. 74, 566–583.

Dineley, D. L. (1979). 'Tectonic setting of Devonian Sedimentation'. In M. R. House, C. T. Scrutton and M. G. Bassett (Eds), *The Devonian System. Spec. Pap. Palaeont. No. 23,* 145–161.

Dineley, D. L. and Elliott, D. K. (1983). 'New species of *Protopteraspis* (Agnatha, Heterostrain) from the Upper Silurian to Lower Devonian of Northwest Territories, Canada'. *J. Paleont.,* Vol. 57, 474–494.

Dineley, D. L. and Loeffler, E. J. (1974). *Ostracoderm faunas of the Delorme and associated Siluro–Devonian formations, North West Territories, Canada. Spec. Pap. Palaeont. No. 18.*

Dineley, D. L. and Loeffler, E. J. (1979). 'Early vertebrates and the Caledonian earth movements'. In A. L. Harris *et al.* (Eds), *The Caledonides of the British Isles Reviewed,* Geological Society, Academic Press, London.

Donovan, D. T. and Jones, E. J. W. (1979). 'Causes of world-wide changes in sea level'. *J. Geol. Soc.,* Vol. 136, 187–192.

Donovan, R. N. (1975). 'Devonian Lacustrine limestones at the margin of the Orcadian Basin, Scotland'. *J. Geol. Soc.,* Vol. 131, 489–510.

Donovan, R. N. (1980). 'Lacustrine cycles, fish ecology and stratigraphic zonations in the Middle Devonian of Caithness'. *Scott. J. Geol.,* Vol. 16, 35–50.

Donovan, R. N. and Foster, R. (1972). 'Subaqueous shrinkage cracks from the Caithness Flagstones Series (Middle Devonian) of northeast Scotland', *J. Sedim. Petrol.,* Vol. 42, 309–317.

Donovan, R. N., Foster, R. J. and Westoll, T. S. (1974). 'A stratigraphical revision of the Old Red Sandstone of north-eastern Caithness'. *Trans. R. Soc. Edinb.,* Vol. 69, 167–201.

Donovan, R. N., Archer, R., Turner, P. and Tarling, D. H. (1976). 'Devonian Palaeogeography of the Orcadian Basin and the Great Glen Fault'. *Nature,* Vol. 259, 550–551.

Dott, R. H. Jr and Batten, R. L. (1971). *Evolution of the Earth,* McGraw-Hill, New York.

Dumestre, A. and Illing, L. U. (1968). 'Middle Devonian Reefs in Spanish Sahara'. In D. H. Oswald (Ed.), *International Symposium on the Devonian System, Calgary, 1967, II,* 333–350.

Dupont, E. (1881). 'Sur l'origine des calcaires dévoniens de la Belgique'. *Bull. Acad. Roy. Belg. Class Sci. Ser. 3,* Vol. 2, 264–280.

Edwards, D. (1980). 'Early Land Floras'. In A. L. Panchen (Ed.), *The Terrestrial Environment and the Origin of Land Vertebrates,* Academic Press, London, pp. 11–38.

Eicher, D. (1976). *Geologic Time,* 2nd edition, Prentice-Hall, London.

Elloy, R. (1972). 'Réflexions sur quelques environnements récifaux du paléozoique'. *Bull. Centre Rech. Pan-SNPA,* Vol. 6, 1–105.

Embry, A. F. and Klovan, J. E. (1971). 'A Late Devonian reef tract on northeastern Banks Island, Northwest Territories'. *Can. Petrol. Geol. Bull.,* Vol. 19, 760–781.

Embry, A. F. and Klovan, J. E. (1976). 'The Middle–Upper Devonian Clastic Wedge of the Franklinian Geosyncline'. *Can. Petrol. Geol. Bull.,* Vol. 24, 489–639.

Erben, H. K. (1962). 'Zur Analyse und Interpretation der rheinischen und hercynischen Magnafacies des Devons'. *Symposiums-Band 2. Int. Arbeits. Silur/ Devon Grenze, Bonn–Bruxelles 1960,* Schweizerbart'sche Verlagsbuchhandlung, Stuttgart, pp. 42–61.

Erben, H. K. (1964). 'Facies developments in the Marine Devonian of the Old World'. *Proc. Ussher Soc.,* Vol. 1, 92–118.

Fahreus, L. E. (1976a). 'Conodontophorid ecology and evolution related to global tectonics'. In C. R. Barnes (Ed.), *Conodont Palaeoecology. Geol. Ass. Canada, Spec. Pap. 15,* 11–26.

Fahreus, L. E. (1976b). 'Possible Early Devonian conodontophorid provincialism'. *Palaeogeogr. Palaeoclimatol. Palaeoecol.,* Vol. 19, 201–217.

Fischer, A. G. (1979). 'Rhythmic Changes in the Outer Earth'. *Geol. Soc. Lond. Newsletter,* Vol. 8, No. 6, 2–3.

Folk, R. L. (1973). 'Evidence for Peritidal Deposition of Devonian Caballos Novaculite, Marathon Basin, Texas'. *Am. Ass. Petrol. Geol. Bull.,* Vol. 57, 702–725.

Folk, R. L. and McBride, E. F. (1976). 'The Caballos Novaculite revisited, Pt. 1; origin of noraculite members'. *J. Sedim. Petrol.,* Vol. 46, 659–669.

Friend, P. F. (1961). 'The Devonian stratigraphy of north and central Vestspitsbergen'. *Proc. Geol. Soc. Yorks.,* Vol. 33, 77–118.

Friend, P. F. (1967). 'Tectonic implications of sedimentation in Spitsbergen and Midland Scotland'. In D. H. Oswald (Ed.), *International Symposium on the Devonian System, Calgary, 1967, II,* 1141–1148.

Friend, P. F. (1968). 'Tectonic Features of Old Red Sandstone Sedimentation in North Atlantic Borders'. In *North Atlantic—Geology and Continental Drift. Am. Ass. Petrol. Geol. Mem. 12,* 703–710.

Friend, P. F. (1973). 'Devonian Stratigraphy of Greenland and Svalbard'. In M. Pitcher (Ed.), *Arctic Geology, Am. Ass. Petrol. Geol., Mem. 19,* 469–470.

Friend, P. F. (1978). 'Distinctive features of some ancient river systems'. In A. D. Miall (Ed.), *Fluvial Sedimentology. Can. Soc. Petrol. Geol. Mem. 5,* 531–542.

Friend, P. F. (1981). 'Devonian sedimentary basins and deep faults of the northernmost Atlantic borderlands'. *Can. Soc. Petrol. Geol. Mem. 7,* 149–165.

Friend, P. F. and House, M. R. (1964). 'The Devonian Period'. In *The Phanerozoic Time-Scale: A Symposium. Qt. J. Geol. Soc. Lond.,* Vol. 120S, 233–236.

Friend, P. F. and Moody-Stuart, M. (1970). 'Carbonate deposition on the river floodplains of the Wood Bay Formation (Devonian) of Spitsbergen'. *Geol. Mag.,* Vol. 107, 181–195.

Friend, P. F., Alexander-Marrack, P. D., Nicholson, J. and Yeats, A. K. (1976a). 'Devonian sediments of East Greenland. I. Introduction, classification of sequences, petrographic notes'. *Medd. om Grønl.,* Vol. 206, No. 1, 1–56.

Friend, P. F., Alexander-Marrack, P. D., Nicholson, J. and Yeats, A. K. (1976b). 'Devonian sediments of East Greenland. II. Sedimentary structures and fossils'. *Medd. om Grønl.,* Vol. 206, No. 2, 1–91.

Fuller, J. G. C. M. and Porter, J. W. (1969). 'Evaporite formations with petroleum reservoirs in Devonian and Mississippian of Alberta, Saskatchewan and North Dakota'. *Am. Ass. Petrol. Geol. Bull.,* Vol. 53, 909–926.

Gale, N. H., Beckinsale, R. D. and Wadge, A. J. (1979). 'Rb–Sr whole rock dating of acid rocks'. *Geochem. J.,* 27–29.

Gardiner, P. R. R. and MacCarthy, I. A. T. (1981). 'The late Palaeozoic evolution of southern Ireland in the context of tectonic basins and their transatlantic significance'. In J. W. Kerr and A. J. Fergusson (Eds), *Geology of the North Atlantic borderlands. Can. Soc. Petrol. Geol. Mem. 7,* 683–725.

Garrels, R. M. and Perry, E. H. (1974). 'Cycling of carbon, sulfur and oxygen through geologic time'. In E. D. Goldberg (Ed.), *The Sea,* Vol. 5, Wiley-Interscience, New York, pp. 303–336.

Gass, I. G. and Geology Course Team (1972). *Science: A Second Level Course: Surface Processes,* S23 – Block A, The Open University Press, Bletchley.

Ginsburg, R. N. (1975). *Tidal Deposits: A Casebook of Recent Examples and Fossil Counterparts,* Springer-Verlag, Berlin.

Gjelberg, J. (1977). 'Facies analysis of the coal-bearing Vaesalstranda Member (Upper Devonian) of Bjornøya'. *Norsk Polarinst. Arbok.* 71–100.

Goldring, R. (1962). 'The trace fossils of the Baggy Beds (Upper Devonian) of north Devon, England'. *Paläont. Z.*, Vol. 36, 232–251.

Goldring, R. (1971). *Shallow-water Sedimentation as illustrated in the Upper Devonian Baggy Beds. Mem. Geol. Soc. Lond. No. 5.*

Goldring, R. (1978). 'Devonian'. In W. S. McKerrow (Ed.), *The Ecology of Fossils*, Duckworth, London, pp. 125–145.

Goldring, R. and Langenstrassen, F. (1979). 'Open shelf and near-shore clastic facies in the Devonian'. In M. R. House, C. T. Scrutton and M. G. Bassett (Eds), *The Devonian System. Spec. Pap. Palaeont. No. 23*, 81–98.

Gonzalez-Bonorino, G. and Middleton, G. V. (1976). 'A Devonian submarine fauna in western Argentina'. *J. Sedim. Petrol.*, Vol. 46, 56–69.

Gordon, W. A. (1975). 'Distribution by latitude of Phanerozoic evaporite deposits'. *J. Geol.*, Vol. 83, 671–684.

Gregor, C. B. (1970). 'Denudation of the continents'. *Nature*, Vol. 228, 273–275.

Greiner, H. (1978). 'Late Devonian facies inter-relationships in bordering areas of the North Atlantic and their palaeogeographic implications'. *Palaeogeogr. Palaeoclimatol. Palaeoecol.*, Vol. 25, 241–263.

Gressly, A. (1838). *Observations géologiques sur le Jura Soleurois: Neue Denkschr. allg. schweiz, Ges. ges. Naturv. 2*, pp. 1–112.

Gross, W. (1950). 'Die paläontologische und stratigraphische Bedeutung der Wirbeltierfaunen des Old Reds und der marinen altpaläozoischen Schichten'. *Abh. Deutsch. Akad. Wiss. Berlin Jahrg. 1949*, No. 1, 1–130.

Hallam, A. (Ed.) (1973). *Atlas of Palaeobiogeography*, Elsevier, Amsterdam.

Hallam, A. (1974). 'Changing patterns of provinciality and diversity of fossil animals in relation to plate tectonics'. *J. Biogeogr.*, Vol. 1, 213–225.

Hallam, A. (1977). 'Secular changes in marine inundation of U.S.S.R. and North America through the Phanerozoic'. *Nature*, Vol. 269, 769–772.

Hallam, A. (1978). *After the Revolution. An Inaugural Lecture, 22 November 1977*, University of Birmingham.

Hallam, A. (1981). *Facies Interpretation and the Stratigraphic Record,* Freeman, Oxford.

Hallam, A. and Bradshaw, M. J. (1979). 'Bituminous shales and oolitic ironstones as indicators of transgressions and regressions'. *J. Geol. Soc. Lond.,* Vol. 136, 157–164.

Halstead, L. B. (1973). 'The Heterostracan Fishes'. *Biol. Rev.*, Vol. 48, 279–332.

Harland, W. B., Cox, A. V., Llewellyn, P. G., Pickton, C. A. G., Smith, A. G. and Walters, R. (1982). *A Geologic Time Scale*, Cambridge University Press.

Hays, J. D. and Pitman, W. C. III (1973). 'Lithospheric plate motions, sea-level changes and climatic and ecological consequences'. *Nature*, Vol. 246, 18–22.

Heckel, P. (1972). 'Recognition of ancient shallow marine environments'. In J. K. Rigby and W. K. Hamblin (Eds), *Recognition of ancient sedimentary environments. Spec. Pub., Soc. Econ. Paleont. Miner.*, Tulsa, No. 16, 226–286.

Heckel, P. (1974). *Carbonate buildups in the geological record: a review. Spec. Pub. Soc. Econ. Paleont. Miner., Tulsa*, No. 18, 90–154.

Heckel, P. H. and Witzke, B. J. (1979). 'Devonian world palaeogeography determined from distribution of carbonates and related lithic palaeoclimatic indicators'. In M. R. House, C. T. Scrutton and M. G. Bassett (Eds), *The Devonian System. Spec. Pap. Palaeont. No. 23*, 145–161.

Hedberg, H. D. (1961). 'Stratigraphic classification and terminology'. *Int. Geol. Congr. 25 (Norden)*, Part 25, 1–38.

Hoeg, O. A. (1942). 'The Downtonian and Devonian flora of Spitsbergen'. *Norges Svalbard-og Ishavs Undersok.*, Vol. 83, 1–228.

Holland, C. H. (1977). '6. Ireland'. In M. R. House *et al.* (Eds), *A correlation of the Devonian rocks in the British Isles. Geol. Soc. Lond. Spec. Rep. No. 8*, 54–65.

Holland, C. H. (1978). 'Stratigraphical classification and all that'. *Lethaia*, Vol. 11, 85–90.

Holland, C. H. (1979). 'Augmentation and decay of the Old Red Sandstone continent: evidence from Ireland'. *Palaeogeogr. Palaeoclimatol. Palaeoecol.*, Vol. 27, 59–66.

Holland, C. H. and Richardson, J. B. (1977). 'The British Isles'. In *The Siluro–Devonian Boundary. I.U.G.S. Series A, No. 5*, 35–44.

Hollard, H. (1968). 'Le dévonien du Maroc et du Sahara nord-occidental'. In D. H. Oswald (Ed.), *International Symposium on the Devonian System, Calgary, 1967, I*, 203–244.

Hou, H., Wang, S., Gao, L., Xian, S., Bai, S., Cao, X., P'an, J., Liao, W. *et al.* (1979). 'The Devonian System of China'. *Papers 2nd All-China Strat. Congr. 1979, Beijing*, Chinese Academy of Geological Sciences, Beijing, pp. 19–22.

House, M. R. (1963). 'Devonian ammonoid successions and facies in Devon and Cornwall'. *Qt. J. Geol. Soc. Lond.*, Vol. 119, 1–27.

House, M. R. (1975a). 'Facies and time in Devonian tropical areas'. *Proc. Yorks. Geol. Soc.*, Vol. 40, 233–288.

House, M. R. (1975b). 'Faunas and time in the marine Devonian'. *Proc. Yorks. Geol. Soc.*, Vol. 40, 305–317.

House, M. R. (1977). '2. Subdivision of the Marine Devonian'. In M. R. House, J. B. Richardson, W. G. Chaloner, J. R. L. Allen, C. H. Holland and T. S. Westoll (Eds), *A correlation of the Devonian rocks in the British Isles. Spec. Rep. No. 8, Geol. Soc. Lond.*, 4–9.

House, M. R. (1979). 'Biostratigraphy of the early Ammonidea'. In M. R. House, C. T. Scrutton and M. G. Bassett (Eds), *The Devonian System. Spec. Pap. Palaeont. No. 23*, 263–280.

House, M. R. and Ziegler, W. (1977). 'The goniatite and conodont sequences in the early Upper Devonian at

Adorf, Germany'. *Geol. Palaeontol.*, Vol. 11, 69–108.

Howard, J. D. (1972). 'Trace Fossils as Criteria for Recognizing Shorelines in the Stratigraphic Record'. In J. K. Rigby and W. K. Hamblin (Eds), *Recognition of Ancient Sedimentary Environments. Spec. Pub. Soc. Econ. Paleont. Miner.*, Tulsa, No. 16, 215–225.

Hsü, K. J. (1972). 'Origin of saline giants: a critical review after the discovery of the Mediterranean Evaporite'. *Earth Sci. Rev.*, Vol. 8, 371–396.

Huang, C. (1978). 'An outline of the tectonic characteristics of China'. *Eclogae Geol. Helv.*, Vol. 71, 611–635.

Hughes, N. F. (Ed.) (1973). *Organisms and Continents through Time. Spec. Pap. Palaeont. No. 12.*

Irwin, M. L. (1965). 'General theory of epeiric clear water sedimentation'. *Bull. Am. Ass. Petrol. Geol.*, Vol. 49, 445–459.

Jaeger, H. (1979). 'Devonian Graptolithina'. In M. R. House (Ed.), *The Devonian System. Spec. Pap. Palaeont. No. 23*, 335–340.

James, N. P. (1977). 'Facies Models. 7. Introduction to Carbonate Facies Models'. *Geosci. Canada*, Vol. 4, 123–125.

Janvier, P. (1977). 'Les Poissons dévoniens de l'Iran central et de l'Afghanistan'. *Mem. Soc. Géol. France*, No. 8, 277–289.

Janvier, P. and Marcoux, J. (1977). 'Les grès rouges de l'Armutgözlek Tepe: leur faune de poissons (Antiarches, Arthrodires et Crossopterygiens) d'âge devonien supérieur (Nappes d'Anatalya, Taurides occidentales, Turquie)'. *Géol. Méditerr.*, Vol. 4, 183–188.

Johnson, J. G. (1971). 'Timing and Coordination of Orogenic, Epeirogenic, and Eustatic Events'. *Geol. Soc. Am. Bull.*, Vol. 82, 3263–3298.

Johnson, J. G. (1979). 'Devonian brachiopod biostratigraphy'. In M. R. House, C. T. Scrutton and M. G. Bassett (Eds), *The Devonian System. Spec. Pap. Palaeont. No. 23*, 291–306.

Johnson, J. G. and Boucot, A. J. (1973). 'Devonian Brachiopods'. In A. Hallam (Ed.), *Atlas of Palaeobiogeography*, Elsevier, Amsterdam, pp. 89–96.

Johnstone, M. H., Jones, P. J., Koop, W. J., Roberts, J., Gilbert Tomlinson, J., Weevers, J. J. and Wells, A. T. (1968). 'Devonian of Central and Western Australia'. In D. H. Oswald (Ed.), *International Symposium on the Devonian System, Calgary, 1967, I*, 599–612.

Kahn, P. G. and Pompea, S. M. (1978). 'Nautiloid growth rhythms and dynamical evolution of the Earth–Moon system'. *Nature*, Vol. 275, 606–611.

Kalinko, M. K. (1974). 'Relation between salt content and oil–gas potential of continents and seas'. *Int. Geol. Rev.*, Vol. 16, 759–768.

Kay, M. (1951). *North American Geosynclines. Geol. Soc. Am., Mem. 48.*

Kazmierczak, J. (1975). 'Colonial Volvocales (Chlorophyta) from the Upper Devonian of Poland and their palaeoenvironmental significance'. *Acta Palaeont. Pol.*, Vol. 20, 73–85.

Keppie, J. D. (1977). *Tectonics of southern Nova Scotia. Nova Scotia Dept. Mines Pap. 77–1*, 1–34.

Kerr, J. W. (1982). 'Evolution of Sedimentary Basins in the Canadian Arctic'. *Phil. Trans. R. Soc., Ser. A*, Vol. 305, 193–203.

Ki, H. C. (1975). 'Unconformity bounded stratigraphic units'. *Geol. Soc. Am. Bull.*, Vol. 86, 1544–1552.

Kinsman, D. J. J. (1966). 'Gypsum and anhydrite of recent age, Trucial Coast, Persian Gulf'. In J. L. Rau (Ed.), *Second Symposium on Salt, Vol. 1*, North Ohio Geological Society, Cleveland, Ohio, pp. 302–326.

Klapper, G. and Johnson, J. G. (1980). 'Endemism and dispersal of Devonian conodonts'. *J. Paleont.*, Vol. 54, 400–455.

Klapper, G. and Ziegler, W. (1979). 'Devonian conodont biostratigraphy'. In M. R. House, C. T. Scrutton and M. G. Bassett (Eds), *The Devonian System. Spec. Pap. Palaeont. No. 23*, 193–198.

Klemme, H. D. (1980). 'Petroleum basins — classifications and characteristics'. *J. Petrol. Geol.*, Vol. 3, 187–207.

Klingspor, A. M. (1969). 'Middle Devonian evaporites of Western Canada'. *Am. Ass. Petrol. Geol. Bull.*, Vol. 53, 927–948.

Krebs, W. (1971). 'Devonian reef limestones of the eastern Rhenisch Schiefergebirge'. In G. Muller (Ed.), *Sedimentology of Parts of Central Europe, 8th Int. Sedimental. Congr. Heidelberg, Guidebook*, 45–81.

Krebs, W. (1974). 'Devonian carbonate complexes of central Europe'. In L. F. Laporte (Ed.), *Reefs in time and space. Spec. Pub. Soc. Econ. Paleont. Miner.*, Tulsa, No. 18, 155–208.

Krebs, W. (1977). 'The Tectonic Evolution of Variscan Meso-Europa. In D. V. Ager and M. Brooks (Eds), *Europe from Crust to Core*, Wiley, London, pp. 119–142.

Krebs, W. (1979). 'Devonian Basinal Facies'. In M. R. House, C. T. Scrutton and M. G. Bassett (Eds), *The Devonian System. Spec. Pap. Palaeont. No. 23*, 125–139.

Krebs, W. and Wachendorf, H. (1973). 'Proterozoic–Palaeozoic geosynclinal and orogenic evolution of central Europe'. *Geol. Soc. Am. Bull.*, Vol. 84, 2611–2630.

Krebs, W. and Wachendorf, H. (1974). 'Faltungskerne im mitteleuropaischen Grundgebirge — Abbilder eines Orogen Diapirismus'. *Neues Jahrb. Geol. Paläont. Abh.*, Vol. 147, 30–60.

Krumbein, W. C. and Sloss, L. L. (1963). *Stratigraphy and Sedimentation*, 2nd edition, Freeman, San Francisco.

Krylova, A. K., Malitch, L. S., Menner, V. V., Obrutchev, D. V. and Fradkin, G. S. (1968). 'Devonian of the Siberian Platform'. In D. H. Oswald (Ed.), *International Symposium on the Devonian System, Calgary, 1967, I*, 473–482.

Krynine, P. D. (1942). 'Differential sedimentation and its products during one complete geosynclinal cycle'.

Proc. 1st Pan. Am. Congr. Mining Eng. Geol., Pt. 1, Vol. 2, 537–560.

Kuendig, E. (1959). 'Eu-geosynclines as potential oil habitats'. *Proc. 5th World Oil Congr.*, Sect. 1, 1–13.

Lambert, J. K. (1971). 'The pre-Pleistocene Phanerozoic time scale—a review'. In W. B. Harland and E. H. Francis (Eds), *Phanerozoic Time Scale Supplement. Geol. Soc. Lond. Spec. Rep. No. 5*, 9–31.

Laporte, L. F. (1967a). 'Carbonate deposition near mean sea level and resultant facies mosaic: Manlius Formation (Lower Devonian) of New York State'. *Am. Ass. Petrol. Geol. Bull.*, Vol. 51, 73–101.

Laporte, L. F. (1967b). 'Recognition of Transgressive Carbonate Sequence within Epeiric Sea: Heldenberg Group (Lower Devonian) of New York State'. *Am. Ass. Petrol. Geol. Bull.*, Vol. 51, 473.

Laporte, L. F. (1969). *Depositional environments in carbonate rocks. Soc. Econ. Paleontol. Mineral. Spec. Pub. 14.*

Laporte, L. F. (1975). 'Carbonate Tidal-Flat Deposits of the Early Devonian Manlius Formation of New York State'. In R. N. Ginsburg (Ed.), *Tidal Deposits*, Springer-Verlag, Berlin, pp. 243–250.

Laporte, L. F. (1979). *Ancient Environments*, 2nd edition, Foundations of Earth Science Series, Prentice-Hall, London.

Laseron, C. (1969). *Ancient Australia: The Story of its Past Geography and Life*, Angus and Robertson, Sydney.

Lecompte, M. (1970). 'Die Riffe in Devon der Ardennen und ihre Bildungsbedingungen'. *Geol. Palaeontol.*, Vol. 4, 25–71.

Leeder, M. R. (1976). 'Sedimentary facies and the origins of basin subsidence along the northern margin of the supposed Hercynian Ocean'. *Tectonophysics*, Vol. 36, 167–179.

Lefort, J. P. (1979). 'Iberian–American arc and Hercynian orogeny in Western Europe'. *Geology*, Vol. 7, 384–388.

Legrande, P. H. (1968). 'Le dévonien du Sahara algérien'. In D. H. Oswald (Ed.), *International Symposium on the Devonian System, Calgary, 1967, I*, 254–284.

Lobanowski, H. and Przybylowicz, T. (1979). 'Tidal flat and flood plain deposits in the Lower Devonian of the western Lublin Uplands (after the boreholes Pionki 1 and Pionki 4)'. *Acta Geol. Polon.*, Vol. 29, 383–406.

Lütke, F. (1979). 'Biostratigraphical significance of the Devonian Dacryoconarida'. In M. R. House, C. T. Scrutton and M. G. Bassett (Eds), *The Devonian System. Spec. Pap. Palaeont. No. 23*, 281–290.

Ma, T. Y. H. (1933). 'On the seasonal change of growth in some Palaeozoic corals'. *Proc. Imp. Akad. Tokyo*, Vol. 9, 407–409.

Ma, T. Y. H. (1934). 'On the seasonal change of growth in a reef coral, *Favia speciosa* (Dana), and the water-temperature of the Japanese seas during the latest geological times'. *Proc. Imp. Acad. Tokyo*, Vol. 10, 353–356.

Maack, R. (1957). 'Über Vereisungsperioden und Vereisungsspuren in Brasilien'. *Geol. Rundsch.*, Vol. 55, 547–595.

Maack, R. (1960). 'Zur Paläogeographie des Gondwanalandes in neuer Beitrag zur Eiszeit des Devons'. *Geol. Jahrb.*, Vol. 25, 1–31.

Mabesoone, J. M., Fulfaro, V. J. and Sugio, K. (1981). 'Phanerozoic sedimentary sequences of the South American platform'. *Earth Sci. Rev.*, Vol. 17, 46–67.

Maiklem, W. R. (1971). 'Evaporative drawdown—a mechanism for water-level lowering and diagenesis in the Elk Point Basin'. *Can. Ass. Petrol. Geol. Bull.*, Vol. 19, 487–503.

Mallory, W. W., Mudge, M. R., Swanson, V. E., Stone, D. S. and Lumb, W. E. (Eds) (1972). *Geologic Atlas of the Rocky Mountain Region, United States of America*, Rocky Mt. Ass. Geol., Denver, Colorado.

Malzahn, E. (1957). 'Devonisches Glazial im Staate Piani (Brasilien), ein neuer Beitrag zur Eiszeit des Devons'. *Beih. Geol. Jahrb.*, Vol. 25, 1–31.

Martinsson, A. (1967). 'The succession and correlation of ostracode faunas in the Silurian of Gotland'. *Geol. Foren. Stockholm Forh.*, Vol. 89, 66–83.

Martinsson, A. (1977). 'Palaeoscope ostracodes'. In D. J. McLaren (Ed.), *The Silurian–Devonian Boundary. I.U.G.S. Series A, No. 5*, 327–332.

McCave, I. N. (1968). 'Shallow and marginal marine sediments associated with the Catskill Complex in the Middle Devonian of New York'. *Geol. Soc. Am. Spec. Pap. 106*, 75–108.

McCave, I. N. (1973). 'The sedimentology of a transgression: Portland Point and Cooksburg Members (Middle Devonian), New York State'. *J. Sedim. Petrol.*, Vol. 43, 484–504.

McElhinney, M. W. (1978). 'The Magnetic Polarity Time Scale: Prospects and Possibilities in Magnetostratigraphy'. In *Contributions to the Geologic Time Scale. Am. Ass. Petrol. Geol. Studies in Geology No. 6*, 57–65.

McGhee, G. R., Jr and Sutton, R. G. (1981). 'Late Devonian marine ecology and zoogeography of the central Appalachians and New York'. *Lethaia*, Vol. 14, 27–43.

McGregor, D. C. (1977). 'Lower and Middle Devonian spores of Eastern Gaspé, Canada. II Biostratigraphy'. *Palaeontographica*, Vol. 163B, 111–142.

McGregor, D. C. (1979). 'Spores in Devonian stratigraphical correlation'. In M. R. House, C. T. Scrutton and M. G. Bassett (Eds), *The Devonian System. Spec. Pap. Palaeont. No. 23*, 163–184.

McKenzie, D. (1978). 'Some remarks on the development of sedimentary basins'. *Earth Planet. Sci. Letters*, Vol. 40, 25–32.

McKerrow, W. S. (Ed.) (1978). *The Ecology of Fossils*, Duckworth, London.

McKerrow, W. S., Lambert, R. St J. and Chamberlain, V. E. (1980). 'The Ordovician, Silurian and Devonian time scales'. *Earth Planet. Sci. Letters*, Vol. 51, 1–8.

McLaren, D. J. (1970). 'Presidential Address: Time, Life and Boundaries'. *J. Paleont.*, Vol. 44, 801–815.

McLaren, D. J. (1977). *The Siluro–Devonian Boundary. I.U.G.S. Series A, No. 5.*

McPherson, J. G. (1979). 'Calcrete (caliche) palaeosols in fluvial red beds of the Aztec Siltstone (Upper Devonian), southern Victoria Land, Antarctica'. *Sedim. Geol.*, Vol. 22, 267–285.

Meer-Mohr, C. G. van der (1975). 'The Palaeozoic strata near Moeche in Galicia, NW Spain'. *Leidse Geol. Meded.*, Vol. 49, 487–497.

Miall, A. D. (1970a). 'Devonian Alluvial Fans, Prince of Wales Island, Arctic Canada'. *J. Sedim. Petrol.*, Vol. 40, 556–571.

Miall, A. D. (1970b). 'Continental marine transition in the Devonian of Prince of Wales Island, Northwest Territories'. *Can. J. Earth Sci.*, Vol. 7, 125–144.

Miall, A. D. (1973). 'Markov chain analysis applied to an ancient alluvial plain succession'. *Sedimentology*, Vol. 20, 347–364.

Miall, A. D. (Ed.) (1981). *Sedimentation and Tectonics in Alluvial Basins. Geol. Ass. Canada, Spec. Pap. 23.*

Middlemiss, F. A., Rawson, P. F. and Newall, G. (Eds) (1971). *Faunal Provinces in Space and Time. Geol. J. Special Issue No. 4*, Seel House Press, Liverpool.

Middleton, G. V. (1978). 'Facies'. In R. W. Fairbridge and J. Bourgeois (Eds), *Encyclopedia of Sedimentology*, Dowden, Hutchinson and Ross, New York, pp. 323–325.

Middleton, V. W. (1973). 'Johannes Walther's Law of the Correlation of facies'. *Geol. Soc. Am. Bull.*, Vol. 84, 979–988.

Miller, A. (1859). *The Testimony of the Rocks*, Constable, Edinburgh.

Mitchell, A. H. G. and Reading, H. G. (1969). 'Continental margins, geosynclines and ocean floor spreading'. *J. Geol.*, Vol. 77, 629–646.

Mitchell, A. H. G. and Reading H. G. (1978). 'Sedimentation and Tectonics'. In H. G. Reading (Ed.), *Sedimentary Environments and Facies*, Blackwell, Oxford, pp. 439–476.

Morel, P. and Irving, E. (1978). 'Tentative palaeocontinental maps for the early Phanerozoic and Proterozoic'. *J. Geol.*, Vol. 86, 535–561.

Morton, D. J. (1979). 'Palaeogeographical evolution of the Lower Old Red Sandstone basin in the western Midland Valley'. *Scott. J. Geol.*, Vol. 15, 97–116.

Mountjoy, E. W. (1975). 'Intertidal and supratidal deposits within isolated Devonian build-ups, Alberta'. In R. N. Ginsburg (Ed.), *Tidal Deposits*, Springer-Verlag, New York, pp. 387–395.

Mountjoy, E. W. and Riding, R. (1981). 'Foreslope stromatoporoid–renalcid bioherm with evidence of early cementation'. *Sedimentology*, Vol. 28, 299–319.

Murphy, M. A. (1977). 'On time-stratigraphic units'. *J. Paleont.*, Vol. 51, 213–219.

Murphy, M. A., Berry, W. B. N. and Sandberg, C. A.

(Eds) (1977). *Western North America: Devonian. Univ. Calif. Riverside Campus Mus. Contrib. 4*, 248.

Mutti, E. and Ricci-Lucchi, F. (1972). *Le torbiditi dell' Appennino settentrionale: introduzione all' analisi di facies. Mem. Soc. Geol. Ital. 11*, 161–199.

Nalivkin, D. V. (1973a). *Geology of the U.S.S.R.* (translated by N. Rast), Oliver and Boyd, Edinburgh.

Nalivkin, D. V. (Ed.) (1973b). 'Stratigrafiya nizhnego i srednego Devona'. *Trudy III Mezh. Simp. po Granitse Siluri i Devona, 2*, Akademia Nauk SSSR, Leningrad.

Narbonne, G., Gibling, M. R. and Jones, B. (1979). '*Polarichnus*, a new trace fossil from Siluro–Devonian strata of Arctic Canada'. *J. Paleont.*, Vol. 53, 133–141.

Nelson, G. and Rosen, D. E. (Eds) (1981). *Vicariance Biogeography: A Critique*, Columbia University Press, New York.

Nilson, T. H. (1968). 'The Relationship of Sedimentation to Tectonics in the Solund Devonian District of southwestern Norway'. *Norges Geol. Undersøk.*, Vol. 259, 1–108.

Nilson, T. H. (1973). 'Devonian (Old Red Sandstone) Sedimentation and Tectonics of Norway'. In M. G. Pitcher (Ed.), *Arctic Geology. Am. Ass. Petrol. Geol. Mem. 19*, 471–481.

North, F. K. (1979). 'Episodes of Source-Sediment Deposition (1)'. *J. Petrol. Geol.*, Vol. 2, 199–218.

North, F. K. (1980). 'The Episodes in Individual Close-up. (Pt. 2 of Episodes of Source-Sediment Deposition)'. *J. Petrol. Geol.*, Vol. 2, 323–338.

Odin, G. S. (Ed.) (1982). *Numerical Dating in Stratigraphy*. Wiley International Science Publication, Wiley, Chichester (2 volumes).

Oliver, W. A., Jr (1973). 'Devonian coral endemism in eastern North America and its bearing on palaeogeography'. In N. Hughes (Ed.), *Organisms and continents through time. Spec. Pap. Palaeont. No. 12*, 319–320.

Oliver, W. A., Jr (1976). 'Presidential address: Biogeography of Devonian Rugose Corals'. *J. Paleont.*, Vol. 50, 365–373.

Oliver, W. A., Jr (1977). 'Biogeography of late Silurian and Devonian rugose corals'. *Palaeogeogr. Palaeoclimatol. Palaeoecol.*, Vol. 22, 85–135.

Oliver, W. A., Jr (1980). 'Corals in the Malvinokaffric Realm'. *Münster. Forsch. Geol. Paläont.*, Vol. 52, 13–27.

Ollier, C. D. (1981). *Geomorphology Texts 6, Tectonics and Landforms*, Longmans, London, pp. 257–267.

Ormiston, A. (1972). 'Lower and Middle Devonian Trilobite Provinces in North America'. *Proc. 23rd Int. Geol. Congr., Montreal*, Sect. 7, p. 594.

Oswald, D. H. (Ed.) (1968). *International Symposium on the Devonian System, Calgary 1967*, I and II, Alberta Soc. Petrol. Geol., Calgary.

Ovenshine, A. T. (1975). 'Tidal Origin of parts of the Karheen Formation (Lower Devonian), Southeastern Alaska'. In R. N. Ginsburg (Ed.), *Tidal Deposits*, Springer-Verlag, New York, pp. 127–134.

Owen, H. G. (1976). 'Continental displacement and

expansion of the Earth during the Mesozoic and Cenozoic'. *Phil. Trans. R. Soc., (Phys. Sci.)*, Vol. 281, 233–291.

Padgett, G., Ehrlich, R. and Moody, M. (1977). 'Submarine debris flow deposits in an extensional setting, Upper Devonian of western Morocco'. *J. Sedim. Petrol.*, Vol. 47, 811–818.

Palmer, A. R. (1965). 'Biomere–a new kind of stratigraphic unit'. *J. Paleont.*, Vol. 39, 149–153.

P'an, K. (1980). 'Devonian Antiarch Biostratigraphy of China'. *Geol. Mag.*, Vol. 118, 69–75.

Petrunkevitch, A. (1953). *Palaeozoic and Mesozoic Arachnida of Europe*. Geol. Soc. Am. Mem. 53.

Pettijohn, F. (1975). *Sedimentary Rocks*, 3rd edition, Harper and Row, New York.

Playford, P. E. (1968). 'Devonian reef complexes in the northern Canning Basin, Western Australia'. In D. H. Oswald (Ed.), *International Symposium on the Devonian System, Calgary, 1967, II*, 351–364.

Playford, P. E. (1980). 'Devonian "Gt. Barrier Reef" of Canning Basin, Western Australia'. *Am. Ass. Petrol. Geol. Bull.*, Vol. 64, 814–840.

Playford, P. E. and Lowry, D. C. (1966). *Devonian reef complexes of the Canning basin, Western Australia*. Geol. Surv. W. Australia Bull. 118.

Poole, W. H., Sanford, B. V., Williams, H. and Kelley, D. G. (1970). 'Chart II, Geotectonic correlation chart for southeastern Canada'. In R. J. W. Douglas (Ed.), *Geology and Economic Minerals of Canada*, Econ. Geol. Rept. No. 1.

Predtechensky, N. N. and Krasnov, V. I. (1967). 'Devonian of intermontane depressions in the Altai Sayan Region'. In D. H. Oswald (Ed.), *International symposium on the Devonian System, Calgary, 1976, I*, 463–472.

Pruvost, P. (1956). *Lexique stratigraphique international. Vol. 1. Europe*, C.N.R.S., Paris, Fasc. 4a and Fasc. 5.

Rau, A. and Tongiorgi, M. (1981). 'Some problems regarding the Palaeozoic paleogeography in Mediterranean Western Europe'. *J. Geol.*, Vol. 89, 663–673.

Raup, D. M. (1972). 'Taxonomic diversity during the Phanerozoic'. *Science*, Vol. 117, 1065–1071.

Raup, D. M. (1976a). 'Species diversity in the Phanerozoic: a tabulation'. *Paleobiol.*, Vol. 2, 279–288.

Raup, D. M. (1976b). 'Species diversity in the Phanerozoic: an interpretation'. *Paleobiol.*, Vol. 2, 289–297.

Read, J. F. (1973). 'Paleo-environments and paleogeography, Pillara Formation (Devonian), Western Australia'. *Can. Petrol. Geol. Bull.*, Vol. 21, 344–394.

Read, H. H. and Watson, J. (1975). *Introduction to Geology*, Macmillan, London (2 volumes).

Reading, H. G. (Ed.) (1978). *Sedimentary Environments and Facies*, Blackwell, Oxford.

Richardson, J. B. (1974). 'The stratigraphic utilization of some Silurian and Devonian miospore species in the northern hemisphere: an attempt at a synthesis'. *Int. Symposium Belgian Micropaleont. Limits. Pub. No. 9*, 1–13.

Rickard, L. V. (1975). *Correlation of the Silurian and Devonian Rocks in New York State. N.Y. State Mus. Sci. Service, Map and Chart Series No. 24*, Albany, N.Y.

Riding, R. (1979). 'Devonian calcareous algae'. In M. R. House, C. T. Scrutton and M. G. Bassett (Eds), *The Devonian System. Spec. Pap. Palaeont. No. 23*, 141–144.

Robinson, P. L. (1973). 'Palaeoclimatology and continental drift'. In D. H. Tarling and S. K. Runcorn (Eds), *Implications of Continental Drift for the Earth Sciences 1*, pp. 451–476.

Rolfe, W. I. D. (1980). 'Early Invertebrate Terrestrial Faunas'. In A. L. Panchen (Ed.), *The Terrestrial Environment and the Origin of Land Vertebrates*, Academic Press, London, pp. 11–38.

Ronov, A. B. and Khain, V. E. (1954). 'Devonian lithographical formations of the world'. *Sov. Geol.*, Vol. 41, 46–76 (in Russian).

Ronov, A. B., Khain, V. E., Balukovsky, A. N. and Seslavinsky, K. B. (1980). 'Quantitative analysis of Phanerozoic sedimentation'. *Sedim. Geol.*, Vol. 25, 311–325.

Rosenberg, G. D. and Runcorn, S. K. (1975). *Growth Rhythms and the History of the Earth's Rotation*, Wiley, London.

Ross, C. A. (Ed.) (1974). *Palaeogeographic Provinces and Provinciality*. Spec. Pub. Soc. Econ. Paleont. Miner., Tulsa, No. 21.

Rudwick, M. J. S. (1979). 'The Devonian: a system born from conflict'. In M. R. House, C. T. Scrutton and M. G. Bassett (Eds), *The Devonian System. Spec. Pap. Palaeont. No. 23*, 9–22.

Runcorn, S. K. (1964). 'Changes in the Earth's moment of inertia'. *Nature*, Vol. 204, 823–825.

Rust, B. R. (1981). 'Alluvial deposits and tectonic style: Devonian and Carboniferous successions in eastern Gaspe'. In A. D. Miall (Ed.), *Sedimentation and Tectonics in Alluvial Basins. Geol. Assoc. Canada Spec. Pap. 23*, 49–76.

Rzhonsnitskaya, M. A. (1968). 'Devonian of the U.S.S.R.'. In D. H. Oswald (Ed.), *International Symposium on the Devonian System, Calgary, 1967, I*, 331–348.

Sandberg, C. A. (1976). 'Conodont biofacies of Late Devonian *Polygnathus styriacus* zone in western United States'. In C. R. Barnes (Ed.), *Conodont Paleoecology. Geol. Ass. Canada Spec. Pap. No. 15*, 171–186.

Sandberg, C. A. and Ziegler, W. (1979). 'Toxonomy and biofacies of important conodonts of Late Devonian *styriacus* Zone, United States and Germany'. *Geol. Palaeontol.*, Vol. 13, 173–212.

Savage, N. M., Perry, D. G. and Boucot, A. J. (1979). 'A Quantitative Analysis of Lower Devonian Brachiopod Distribution'. In J. Gray and A. J. Boucot (Eds), *Historical Biogeography, Plate Tectonics and the Changing Environment*, Oregon State Univ. Press, pp. 169–200.

Schluger, P. R. (1976a). 'Stratigraphy and Sedimentary Environments of the Devonian Perry Formation, New Brunswick, Canada, and Maine, U.S.A.'. *Geol. Soc. Am. Bull.*, Vol. 84, 2533–2548.

Schluger, P. R. (1976b). 'Petrology and origin of the red beds of the Perry Formation, New Brunswick, Canada and Maine, U.S.A.'. *J. Sedim. Petrol.*, Vol. 46, 22–37.

Schmalz, R. F. (1969). 'Deep water evaporite depositions: a genetic model'. *Am. Ass. Petrol. Geol. Bull.*, Vol. 53, 798–823.

Schuchert, C. (1915). A Text-book of Geology by Pirsson and Schuchert, 1st edition, Wiley, New York (2 volumes).

Schuchert, C. (1923). 'Sites and natures of the North-American geosynclines'. *Geol. Soc. Am. Bull.*, Vol. 34, 151–260.

Scotese, C. R., Bambach, R. K., Barton, C., Van Der Voo, R. and Ziegler, A. M. (1979). 'Paleozoic Base Maps'. *J. Geol.*, Vol. 87, 217–277.

Scruton, P. C. (1960). 'Delta building and the deltaic sequence'. In F. P. Shepard, F. B. Phleger and T. H. Van Andel (Eds), *Recent sediments, northwest Gulf of Mexico. Am. Ass. Petrol. Geol.*, 82–102.

Scrutton, C. T. (1965). 'Periodicity in Devonian coral growth'. *Palaeontology*, Vol. 7, 552–558.

Scrutton, C. T. (1978). 'Periodic Growth Features in Fossil Organisms and the length of the Day and Month'. In P. Broche and J. Sundermann (Eds), *Tidal Friction and the Earth's Rotation*, Springer-Verlag, Berlin, pp. 154–196.

Selly, R. C. (1970). *Ancient Sedimentary Environments*, Chapman and Hall, London.

Sepkoski, J. J., Jr (1976). 'Species diversity in the Phanerozoic: species–area effects'. *Paleobiol.*, Vol. 2, 289–303.

Shaw, A. B. (1964). *Time in Stratigraphy*, McGraw-Hill, New York.

Shearman, D. J. and Fuller, J. G. C. M. (1969). 'Anhydrite diagenesis, calcitization, and organic laminites, Winnipegosis formation, Middle Devonian, Saskatchewan'. *Can. Petrol. Geol. Bull.*, Vol. 17, 496–525.

Shirley, J. (1938). 'Some aspects of the Siluro–Devonian Boundary Problem'. *Geol. Mag.*, Vol. 75, 353–362.

Shirley, J. (1963). 'The Distribution of Lower Devonian Faunas'. In A. E. M. Nairn (Ed.), *Problems in Palaeoclimatology*, Interscience Publishers, London, pp. 255–261, 297–299.

Simpson, G. G. (1960). 'Notes on the measurement of faunal resemblance'. *Am. J. Sci.*, Vol. 258, 300–311.

Sloss, L. L. (1963). 'Sequences in the cratonic interior of North America'. *Geol. Soc. Am. Bull.*, Vol. 74, 93–114.

Sloss, L. L. (1972). 'Synchrony of Phanerozoic Sedimentary–Tectonic events of the North American Craton and the Russian Platform'. *Proc. 24th Int. Geol. Congr.*, Sect. 6, 24–32.

Sloss, L. L. and Speed, R. C. (1974). 'Relationships of cratonic and continental-margin tectonic episodes'. In W. R. Dickinson (Ed.), *Tectonics and Sedimentation.*

Spec. Pub. Soc. Econ. Paleont. Miner., *Tulsa*, No. 22, 98–119.

Smirnov, V. G. (1973). 'Tectonic regime and hydrodynamics of marine waters as factors in deposition of Domanik sediments in Tartaria'. *Dokl. Akad. Nauk SSSR*, Vol. 211, 456–459 (in Russian).

Smith, N. D. (1970). 'The braided stream depositional environment: comparison of the Platte River with some Silurian clastic rocks'. *Geol. Soc. Am. Bull.*, Vol. 81, 2993–3014.

Smith, A. G., Hurley, A. M. and Briden, J. C. (1981). *Phanerozoic Paleocontinental World Maps*, Cambridge University Press.

Sokolov, B. S. (1978). *Geology of the Northwest Russian Platform*, Zinatre, Riga.

Sokolov, B. S. and Rzhonsnitskaya, M. A. (1982). *Biostratigraphy of Lower and Middle Devonian Boundary Deposits*. Academy of Sciences of the U.S.S.R., VSEGEI, Nauka, Leningrad (in Russian with English abstracts).

Sokolov, B. S., Lyarskaya, L. A. and Sauvaitova, L. S. (1981). *Devonian and Carboniferous of the Prebaltic Region*, Zinatne, Riga (in Russian).

Solle, G. (1976). 'Oberes Unter und unteres Mitteldevon einer typischen Geosynklinal-Folge im südlichen Rheinischen Schieforgebirge. Die Olkenbacher Mulde'. *Geol. Abh. Hessen*, Vol. 74, 1–264.

Stauffer, K. W. (1968). 'Silurian–Devonian reef complex near Nowshera, West Pakistan'. *Geol. Soc. Am. Bull.*, Vol. 79, 1331–1350.

Steel, R. J. (1976). 'Devonian basins of Western Norway—Sedimentary response to tectonism and to varying tectonic context'. *Tectonophysics*, Vol. 36, 207–224.

Steel, R. J. and Aasheim, S. M. (1978). 'Alluvial sand deposition in a rapidly subsiding basin (Devonian, Norway). In A. D. Miall (Ed.), *Fluvial Sedimentology. Can. Soc. Petrol. Geol. Mem. 5*, 385–412.

Steel, R. J., Maehle, S., Nilsen, H., Røe, S. L. and Spinnangr, Å. (1977). 'Coarsening-upward cycles in the alluvium of Hornelen Basin (Devonian) Norway: Sedimentary response to tectonic events'. *Geol. Soc. Am. Bull.*, Vol. 88, 1124–1134.

Steiger, R. H. and Jäger, E. (1977). 'Subcommission on Geological Time'. *Earth Planet. Sci. Letters*, Vol. 36, 359.

Steiner, J. (1973). 'Possible galactic causes for synchronous sedimentation sequences of the North American and eastern European cratons'. *Geology*, Vol. 1, 89–92.

Størmer, L. (1977). 'Arthropod Invasion of Land During Late Silurian and Devonian Times'. *Science*, Vol. 197, 1362–1364.

Sutton, R. G., Bowen, Z. P. and McAlester, A. L. (1970). 'Marine shelf environments of the Upper Devonian Sonyea Group of New York'. *Geol. Soc. Am. Bull.*, Vol. 88, 2975–2992.

Tarling, D. H. (1980). 'Upper Palaeozoic Continental

Distributions Based on Palaeomagnetic Studies'. In A. L. Panchen (Ed.), *The Terrestrial Environment and the Origin of Land Vertebrates*, Academic Press, London, pp. 11–38.

Tarlo, L. B. H. (1965). 'Psammosteiformes (Agnatha)—a review with descriptions of new material from the Lower Devonian of Poland. I. General Part'. *Palaeontol. Pol.*, Vol. 13, 1–135.

Thayer, C. W. (1974). 'Marine palaeoecology in the Upper Devonian of New York'. *Lethaia*, Vol. 7, 121–155.

Thorsteinsson, R. and Tozer, E. T. (1968). 'X. Geology of the Arctic Archipelago'. In J. R. W. Douglas (Ed.), *Geology and Economic Minerals of Canada. Geol. Surv. Canada. Econ. Geol. Rep. No. 1*, 547–590.

Till, R. (1978). 'Arid Shorelines and Evaporites'. In H. G. Reading (Ed.), *Sedimentary Environments and Facies*, Blackwell, Oxford, pp. 178–206.

Trettin, H. P. (1972). 'The Innuitian Province'. In R. A. Price and R. J. W. Douglas (Eds), *Variations in tectonic styles in Canada. Geol. Ass. Canada, Spec. Pap. 11*, 83–179.

Trettin, H. P. (1973). 'Early Palaeozoic evolution of northern parts of Canadian Arctic Archipelago'. In M. Pitcher (Ed.), *Arctic Geology. Am. Ass. Petrol. Geol. Mem. 19*, 57–75.

Trewin, N. H. (1976). '*Isopodichnus* in a trace fossil assemblage from the Old Red Sandstone'. *Lethaia*, Vol. 9, 29–37.

Tsien, H. H. (1977). 'Morphology and development of Devonian reefs and reef complexes in Belgium'. *3rd Int. Coral Reef Sympos., Univ. Miami, Florida*, Vol. 2, 191–200.

Tsien, H. H. (1979). 'Paleoecology of algal-bearing facies in the Devonian (Couvinian to Frasnian) reef complexes of Belgium'. *Palaeogeogr. Palaeoclimatol. Palaeoecol.*, Vol. 27, 103–127.

Tucker, M. E. (1969). 'Crinoidal turbidites from the Devonian of Cornwall and their palaeogeographic significance'. *Sedimentology*, Vol. 13, 281–290.

Tucker, M. E. (1973). 'Sedimentology and diagenesis of Devonian pelagic limestones (Cephalopodenkalk) and associated sediments in the Rhenohercynian geosyncline, West Germany'. *Neues Jb. Geol. Paläont. Abh.*, Vol. 142, 320–350.

Tucker, M. E. (1974). 'Sedimentology of Palaeozoic pelagic limestones: the Devonian Griotte (southern France) and Cephalopodenkalk (Germany). In K. J. Hsü and H. C. Jenkyns (Eds), *Pelagic sediments on land and under the sea. Spec. Pub. Int. Ass. Sediment. 1*, 71–92.

Tucker, M. E. and Kendall, A. C. (1973). 'The diagenesis and low-grade metamorphism of Devonian styliolinid-rich pelagic carbonates from West Germany: possible analogues of Recent pteropod oozes'. *J. Sedim. Petrol.*, Vol. 43, 672–687.

Turner, P. (1980). *Developments in Sedimentology, 29: Continental Red Beds*, Elsevier, Amsterdam.

Turner, S. (1971). 'Thelodonts and the Siluro–Devonian Boundary'. *J. Geol. Soc. Lond.*, Vol. 127, 632–635.

Turner, S. and Tarling, D. H. (1982). 'Thelodont and other agnathan distributions as tests of Lower Palaeozoic continental reconstructions'. *Palaeogeogr. Palaeoclimatol. Palaeoecol.*, Vol. 39, 295–311.

Umbgrove, J. H. F. (1947). *The Pulse of the Earth*, Martinus Nijhoff, The Hague, 2nd edition.

Valentine, J. W. (1970). 'How many marine invertebrate fossil species? A new approximation'. *J. Paleont.*, Vol. 44, 410–415.

Valentine, J. W. (1973). *Evolutionary Palaeoecology of the Marine Biosphere*, Prentice-Hall, Englewood Cliffs, New Jersey.

Valentine, J. W. and Moores, E. M. (1970). 'Plate-tectonic regulation of faunal diversity and sea level: a model'. *Nature*, Vol. 228, 657–659.

Valentine, J. W. and Moores, E. M. (1972). 'Global tectonics and the fossil record. *J. Geol.*, Vol. 80, 167–184.

Valentine, J. W. and Moores, E. M. (1974). 'Plate Tectonics and the History of Life in the Oceans'. *Sci. Am.*, April.

Van Der Voo, R., Briden, J. C. and Duff, B. A. (1980). 'Late Precambrian and Palaeozoic palaeomagnetism of the Atlantic-bordering continents'. In *Geology of Europe. Fr. B. R.G.M. Mem. 108*, 202–212.

Van Eysinga, F. W. B. (1975). *Geological Time Table*, 3rd edition, Elsevier, Amsterdam.

Van Hinte, J. E. (1978). *Working Group on Unified Stratigraphic Time-Scale. UTS Circular 1*, 11.

Van Houten, F. B. (Ed.) (1977). *Ancient Continental Deposits. Benchmark Papers in Geology 43*, Dowden, Hutchinson and Ross, Strondsberg, Pennsylvania.

Vinogradov, A. P. (Ed.) (1969). *Atlas of the lithological–palaeogeographical maps of the U.S.S.R. Vol. 2.* Ministry of Geology, Academy of Sciences of U.S.S.R., Moscow.

Vos, R. G. (1981). 'Deltaic sedimentation in the Devonian of western Libya'. *Sedim. Geol.*, Vol. 29, 67–88.

Walcott, R. I. (1970). 'An isostatic origin for basement uplifts'. *Can. J. Earth Sci.*, Vol. 7, 931–937.

Walker, K. R. and Laporte, L. F. (1970). 'Congruent fossil communities from Ordovician and Devonian carbonates of New York'. *J. Paleont.*, Vol. 44, 928–944.

Walker, R. G. (1971). 'Non-deltaic depositional environments in the Catskill clastic wedge (Upper Devonian) of central Pennsylvania'. *Geol. Soc. Am. Bull.*, Vol. 82, 1305–1326.

Walker, R. G. (1975). 'From sedimentary structures to facies models'. In *Depositional Environments as Interpreted from Primary Sedimentary Structures and Stratigraphic Sequences. S.E.P.M. Short Course No. 2*, 163–180.

Walker, R. G. (1979). *Facies and Facies Models. 1. General Introduction. Geoscience Canada, Reprint Series 1*, 1–7.

Walker, R. G. and Harms, J. C. (1971). 'The "Catskill Delta": a prograding muddy shoreline in central Pennsylvania'. *J. Geol.*, Vol. 79, 381–399.

Walker, R. G. and Harms, J. C. (1975). 'Shorelines of Weak Tidal Activity; Upper Devonian Catskill Formation, Central Pennsylvania'. In R. N. Ginsburg (Ed.), *Tidal Deposits*, Springer-Verlag, Berlin, pp. 103–108.

Walker, R. G. and Mutti, E. (1973). 'Turbidite facies and facies associations'. In *Turbidites and Deep Water Sedimentation, S.E.P.M. Short Course*, 119–157.

Walmsley, V. G. (1974). 'The base of the Upper Palaeozoic'. In T. R. Owen (Ed.), *The Upper Palaeozoic and Post Palaeozoic Rocks of Wales*, University of Wales Press, Cardiff, pp. 33–45.

Wang, Y., Yu, C., Ruan, Y. and Wang, C. (1981). 'Subdivision of the Devonian of China'. *Kexue tongbao*, Vol. 26, 639–642.

Webby, B. D. (1965). 'The Middle Devonian marine transgression in North Devon and West Somerset'. *Geol. Mag.*, Vol. 102, 478–488.

Weber, K. J. (1971). 'Sedimentological aspects of oil fields in the Niger Delta'. *Geol. Mijnb.*, Vol. 50, 559–576.

Weller, J. M. (1960). *Stratigraphic Principles and Practices*, Harper and Brothers, New York.

Wells, J. W. (1944). 'Fish remains from the Middle Devonian bone beds of the Cincinnati Arch region'. *Palaeont. Am.*, Vol. III, No. 16, 5–63.

Wells, J. W. (1963). 'Coral growth and geochronometry'. *Nature*, Vol. 197, 948–950.

Wells, J. W. (1970). 'Problems of annual and daily growth rings in corals'. In S. K. Runcorn (Ed.), *Palaeogeophysics*, Academic Press, London, pp. 3–9.

Westoll, T. S. (1977). '7. Northern Britain'. In M. R. House *et al.* (Eds), *A correlation of the Devonian rocks in the British Isles. Geol. Soc. Spec. Rep. No. 8*, 66–92.

Westoll, T. S. (1979). 'Devonian fish biostratigraphy'. In M. R. House, C. T. Scrutton and M. G. Bassett (Eds), *The Devonian System. Spec. Pap. Palaeont. No. 23*, 341–351.

Westoll, T. S., Shirley, J., Dineley, D. L. and Ball, H. W. (1971). 'The Silurian–Devonian Boundary. Written contribution to the paper "A correlation of Silurian rocks in the British Isles"'. *J. Geol. Soc.*, Vol. 127, 285–288.

Wheeler, H. E. (1964). 'Baselevel, lithosphere surface, and time-stratigraphy'. *Geol. Soc. Am. Bull.*, Vol. 75, 599–610.

White, E. I. (1950). 'The vertebrate faunas of the Lower Old Red Sandstone of the Welsh Borders'. *Bull. Brit. Mus. (Nat. Hist.) Geol.*, Vol. 1, 51–67.

Whyte, M. A. (1977). 'Turning points in Phanerozoic history'. *Nature*, Vol. 267, 679–682.

Wilson, J. L. (1975). *Carbonate Facies in Geologic History*, Springer-Verlag, Berlin.

Wise, D. U. (1974). 'Continental Margins, Freeboard and Volumes of Continents and Oceans Through Time'. In C. A. Burk and C. L. Drake (Eds), *The Geology of Continental Margins*, Springer-Verlag, Berlin, pp. 45–58.

Woodrow, D. L., Fletcher, F. W. and Ahrnsbrak, W. F. (1973). 'Paleogeography and Paleoclimate at the Deposition Sites of the Devonian Catskill and Old Red Facies'. *Geol. Soc. Am. Bull.*, Vol. 84, 3051–3064.

Yang, S., P'an, K. and Hou, H. (1981). 'The Devonian System in China'. *Geol. Mag.*, Vol. 118, 113–224.

Yeats, A. K. and Friend, P. F. (1978). Devonian sediments of East Greenland. IV. The western sequences. Kap Kolthoff super-group of the western areas. *Medd. om. Gronl.*, Vol. 206, No. 4.

Young, G. C. (1981). 'Biogeography of Devonian vertebrates'. *Alcheringa*, Vol. 5, 225–243.

Zharkov, M. A. (1981). *History of Paleozoic Salt Accumulation*, Springer-Verlag, Berlin.

Ziegler, A. M., Cocks, L. R. M. and Bambach, R. K. (1968). 'The composition and structure of lower Silurian marine communities'. *Lethaia*, Vol. 1, 1–17.

Ziegler, A. M., Hansen, K. S., Johnson, M. E., Kelly, M. A., Scotese, C. R. and Van Der Voo, R. (1977). 'Silurian continental distributions, paleogeography, climatology and biogeography'. *Tectonophysics*, Vol. 40, 13–51.

Ziegler, A. M., Scotese, C. R., McKerrow, W. S., Johnson, M. E. and Bambach, R. K. (1979). 'Palaeozoic Palaeogeography'. *Ann. Rev. Earth Planet. Sci.*, Vol. 7, 473–502.

Ziegler, W. (1979). 'Historical subdivisions of the Devonian'. In M. R. House, C. T. Scrutton and M. G. Bassett (Eds), *The Devonian System. Spec. Pap. Palaeont. No. 23*, 23–48.

Zonenshain, L. P. (1973). 'The evolution of central Asiatic Geosynclines through sea floor spreading'. *Tectonophysics*, Vol. 19, 213–232.

Author Index

203

Fossil Genera and
Species Index

Subject Index